ENGINEERING HISTORY AND HERITAGE

PROCEEDINGS OF THE SECOND NATIONAL CONGRESS ON CIVIL ENGINEERING HISTORY AND HERITAGE

SPONSORED BY
ASCE Committee on History and Heritage
of American Civil Engineering (CHHACE),

The Boston Society of Civil Engineers
on their 150th Anniversary (1848–1998)

and the
National Park Service

October 17–21, 1998
Boston, Massachusetts

EDITED BY
Jerry R. Rogers

1801 ALEXANDER BELL DRIVE
RESTON, VIRGINIA 20191–4400

Abstract: This proceedings, *Engineering History and Heritage*, contains papers presented at Second National Congress on Civil Engineering History and Heritage, held in conjunction with the ASCE Annual Convention in Boston, Massachusetts, October 18-21, 1998. In order to encourage the development of course material for the teaching of civil engineering history and heritage, this publication presents a variety of information on civil engineering history. Specific landmarks, such as the Moseley Riveted Wrought Iron Arch, are discussed, as are the programs involved in preserving engineering landmarks. In addition, other papers explore the history of underground engineering and construction. Finally, twelve geographic sections of the American Society of Civil Engineers provide information on the history of civil engineering in their areas of the country.

Library of Congress Cataloging-in-Publication Data

National Congress on Civil Engineering History and Heritage (2nd: 1998: Boston, Mass.)
Engineering history and heritage: proceedings of the Second National Congress on Civil Engineering History and Heritage / sponsored by the Committee on History and Heritage of the American Society of Civil Engineers (CHHACE), the Boston Society of Civil Engineers on their 150th anniversary (1848-1998), and National Park Service in conjunction with the ASCE National Convention in Boston, Massachusetts, October 17-21, 1998; edited by Jerry R. Rogers.
p. cm.
Includes bibliographical references and index.
ISBN 0-7844-0394-5
1. Civil engineering–United States–History–Congresses. 2. Civil engineers–United States–Congresses. I. Rogers, Jerry R. II. American Society of Civil Engineers. Committee on History and Heritage of American Civil Engineering. III. Boston Society of Civil Engineers. IV. United States. National Park Service. V. ASCE National Convention (1998: Boston, Mass.) VI. Title.
TA5.N314 1998 98-38924
624'.0973–dc21 CIP

Library of Congress Catalog Card No: 98-38924
ISBN 0-7844-0394-5
Manufactured in the United States of America.

Engineering History and Heritage

Jerry R. Rogers, Ph.D., P.E., Chair: CHHACE, ASCE History Committee,
Editor/Coordinator: Second National Engineering History and Heritage Congress
University of Houston, Houston, Texas 77204

The 1998 National Engineering History and Heritage Congress is co-sponsored by the Boston Society of Civil Engineers (Anni H. Autio, Robert J. LoConte, and David L. Westerling), the National Park Service (Eric DeLony and Robie Lange), and the American Society of Civil Engineers (ASCE) Committee on the History and Heritage of American Civil Engineering (CHHACE) (Augustine J. Fredrich, William P. Johnson, Jr., Alan L. Prasuhn- Past Chair, Edward L. Robinson, Jerry R. Rogers-Chair, Malcomb L. Steinberg, and Jane Howell, Staff Liaison).

Presidents Luther W. Graef and Daniel S. Turner have challenged the ASCE Education Activities Committee and CHHACE to encourage the development of course material for the teaching of civil engineering history and heritage in universities. New engineering history and heritage course material is provided in the 1998 ASCE History Congress Proceedings which followed engineering history information in the 1996 National Symposium on Civil Engineering History and ASCE Proceedings ("**CIVIL ENGINEERING HISTORY**: Civil Engineers Make History") co-sponsored by CHHACE, the National Capital Section- ASCE, and the U.S. Capitol Historical Society. Also, an Engineering History and Landmarks Workshop held July 13, 1998 in Washington, D.C. with ASME, Education Activities Committee- ASCE, the National Park Service, CHHACE-ASCE, and 10+ professional societies and agencies recommended a permanent engineering history coordinating council and regular workshops. Therefore, there are new developments to discover new material for engineering history and heritage.

The Boston Society of Civil Engineers was founded 150 years ago in 1848 and the 150th Anniversary Celebration occurs during the 1998 ASCE Annual Convention in Boston. The Engineering History Field trips are the October 17-18 New England Covered Bridges Tour (Frank Griggs, Jr. and BSCES Historians) and the October 18 Lowell National Industrial Park Tour (Lawrence R. Keats and Robert J. LoConte). The Boston National Engineering History and Heritage Congress Sessions and Coordinators included the Annual ASCE National History and Heritage Breakfast (Speaker: Augustine J. Fredrich: "European Medieval Cathedrals,"), "Three Centuries of Engineering Aesthetics" (Reuben F. Hull, Jr. and Frederick M. Law), "150 Years of Underground Construction" (Ronald C. Hirschfeld), "These Olde Pyramids- the Sphinx and the Pyramids of Egypt" (Robert J. Kachinsky and Robert J. LoConte), "The Boston Society of Civil Engineers 150th Anniversary Dinner at the Phillips Winthrop House and Tour" (Anni H. Autio and David L. Westerling), "National

Engineering Historical Landmarks and History- Parts I and II" and the "Big 12 Sections History Congress" (Jerry R. Rogers).

The first ASCE Sections History Congress is particularly interesting due to excellent papers from ASCE Sections in Los Angeles, Seattle, Inland Empire/Spokane, Colorado, Texas, St. Louis, Tennessee/Nashville (where the ASCE 150th Anniversary Convention will be held in 2002), National Capital, Maryland, Metropolitan, Buffalo, and Philadelphia.

Special acknowledgments go to the 1998 ASCE National History and Heritage Award winners: Irving Sherman- Los Angeles Section- ASCE and Joe E. Colcord- Seattle Section- ASCE and the Philadelphia Section- ASCE for receiving the 1998 Presidential/CHHACE Outstanding Section Award. Also, special thanks go to Jane Howell, Gayle Fields, Alan L. Prasuhn, and Edward L. Robinson for the revised ASCE history and heritage guidebook. The new, attractive ASCE guidebook has much information for educators, practitioners, and students interested in civil engineering history, heritage and landmarks. **Copies of the new ASCE Guidebook may be obtained from ASCE Washington Office/CHHACE, 1015- 15th Street, N.W., Suite 600, Washington, D.C. 20005-2605.** The Los Angeles Section- ASCE is to be commended for putting together their numerous landmarks with digital photographs suitable for the internet educational media with a special regional presentation August 20 at the Huntington Library. Neal FitzSimons is working with ASCE and Michael Kupferman with Frank Griggs, Jr. to develop the ASCE Landmarks and engineering history timeline on the Internet by the fall of 1998. Also, the Metropolitan Section- ASCE has a most attractive booklet of landmarks in the New York City region. Several ASCE Sections are doing great projects to inform and educate people on civil engineering projects, history and heritage.

The third ASCE Engineering History and Heritage Congress is being planned for the Houston, Texas ASCE Annual Convention in 2001. The 2001 National/ International Engineering History and Heritage Congress and ASCE Proceedings will include international engineering societies and history, more Sections Engineering History Sessions, National Engineering History and Heritage Sessions, and the beginning of the 150th ASCE Anniversary, culminating at the Nashville ASCE Convention Celebration.

Commemorative Print

As a lasting tribute to 150 years of civil engineering accomplishments, the Boston Society of Civil Engineers Section/ASCE commissioned architect Frederick "Fritz" T. Kubitz to produce a 150th Anniversary commemorative pen-and-ink print of a collection of local significant civil engineering works. Of the projects represented in the collage, the Canton Viaduct, built in 1835, and the Moseley Bridge, built in 1864, will be dedicated in 1998 as ASCE National Historic Civil Engineering Landmarks. Copies of the commemorative print are on sale at The Engineering Center. Proceeds from the sale will be donated to The Engineering Center Education Trust.

Canton Viaduct, Canton, MA
1998 National Historic Civil Engineering Landmark
Opened in 1835, the Canton Viaduct has been and will be in continuous service to high speed rail to 1999 and beyond. This 21-arch granite masonry bridge was uniquely designed with hollow spaces between walls, connected by solid buttresses between arches. The slightly curved, functional bridge is 615 feet long, 70 feet high, and 22 feet wide.

Moseley Wrought Iron Arch Bridge, North Andover, MA
1998 National Historic Civil Engineering Landmark
Designed, patented in 1857, and built by Thomas W.H. Moseley in 1864, this arched 96- foot span bridge incorporated for the first time in the United States, the use of riveted wrought iron plates for the triangular- shaped top chord and preceded by years the standard use of wrought iron for bridges.

CREDIT FOR COVER ILLUSTRATION

Phillips Winthrop House
Home of The Engineering Center and BSCES/ASCE
Since 1990, the Phillips Winthrop House, One Walnut Street, has been the home of The Engineering Center and the Boston Society of Civil Engineers Section/ ASCE. The BSCE was founded in 1848 and the 1998 Engineering History and Heritage Congress October 17-21, 1998 in Boston marks the 150th Anniversary Celebration of the BSCE. The Phillips Winthrop House was built in 1804 and was a product of Charles Bullfinch, the leading architect of his time.

Contents

---

† HH1 represents 1998 National Engineering History & Heritage Congress Session: 1.
* Manuscript not available at time of printing.

---

* Manuscript not available at time of printing.

Planning Begins for the 2001 National/International Engineering History and Heritage Congress in Houston, Texas at the ASCE Annual Convention in October 2001.

The Moseley Riveted Wrought Iron Arch

a

National Historic Civil Engineering Landmark

Francis E. Griggs, Jr.[1]

Abstract: The Moseley Wrought Iron Arch, originally built across the North Canal in Lawrence, Massachusetts in 1864 was recently restored by the students, faculty and friends of Merrimack College in North Andover, Massachusetts. It is the oldest riveted wrought iron bridge in the United States and one of the wrought iron bridges in the country. It was named a National Historic Civil Engineering Landmark in 1998.

Introduction

Thomas W. H. Moseley was born in Kentucky on November 28, 1813. He apprenticed "to the first iron furnace built on the Ohio River, known as the Union Furnace" (Moseley 1948) at Hanging Rock near what is now called Ironton, Ohio. The iron making industry in the United States was still in its infancy and was centered in Pennsylvania. By the end of the 1850's, however, there were four mills in the Cincinnati/Covington area producing boiler plate. This opportunity to make cast iron and later wrought iron gave Moseley an introduction not only to iron making, but also to its uses. He knew, for example, of Capt. Delafield's cast iron arch bridge on the National Road at Brownsville, Pennsylvania built in 1839. He had, in fact, superintended the weighing and shipping of the iron used in that bridge. It was the first iron bridge built in the United States and still stands at its original site.

He became interested in bridges in the 1850s writing "almost every individual who, as its engineer, has made ten miles of road, has at one time or another conceived a new plan of bridge; for of all the troubles which beset

[1] Director of Quality Assurance, Clough, Harbour & Associates LLP, lll Winners Cricle, Albany, NY 12205

an engineer in constructing and operating a road, its Bridging is the greatest." (Moseley 1863) He had been exposed to wooden bridges built by Lewis Wernwag and Theodore Burr who used arches extensively. He apparently read the English and American technical journals of the day, as he was familiar with Robert Stephenson's Britannia Bridge made of wrought iron plates riveted together into a huge rectangular tube. He considered it to be "a gigantic monument to the brute force of labor and money." (Moseley 1863) He also knew of Thomas Telford's iron arch bridges built in England, Scotland and Wales.

In 1853, while building a road in Kentucky, he hit upon an idea of a wrought iron riveted plate tied arch as a solution to the bridge problem. He may have been influenced by Frederick Harback's iron truss which used wrought iron plates riveted together in the shape of a circular tube for the lower tension chord, Harback went on to build several of these bridges on the Cleveland, Columbus and Cincinnati Railroad in the period 1848-50. Moseley was working in this area at the same time and may have seen some of Harback's trusses. However, he wasn't fond of pure trusses, believing that "no bridge should be considered safe without the arch." (Moseley 1863)

Moseley was issued patent no. 16,572, Figure 1, on February 3, 1857 for a bowstring "truss bridge." The bridge was fabricated entirely of wrought iron plate, bar and strap stock.

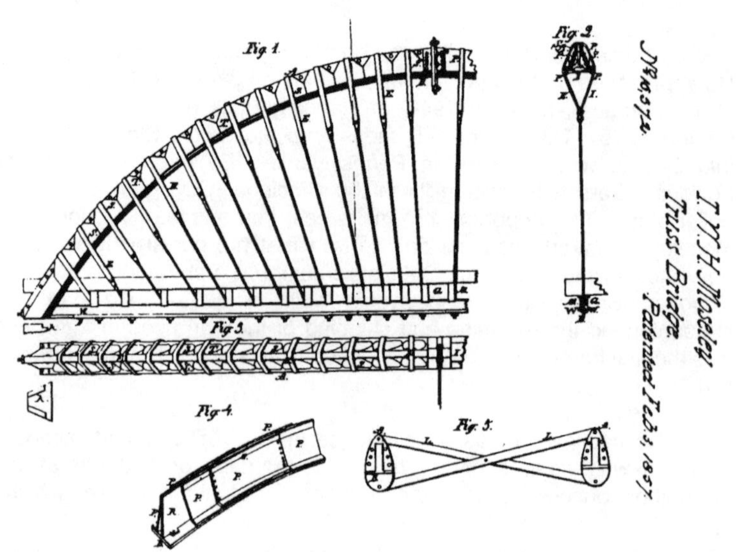

Figure 1 Patent Drawing No. 16,572

After being issued his patent in 1858 he set up a factory in Cincinnati to build his bridges. He advertised that "we are prepared to construct and erect our bridges in every part of the US, the Canada, the Indies, &c. with single spans up to 2000 feet (though in long bridges with single spans the increase in cost is very great,) A bridge of 50 feet or less in span, we can construct in three days time, and when it is on the ground and ready for placing in position, we require but a few hours to remove the old one and place the new one complete in its stead..."(Railroad Record 1857)

He continued to build over 60 of his bridges prior to the Civil War with 17 of them in Ohio, several in Illinois, Indiana, Virginia, and 8 in Kentucky. One of his bridges was a 59' long aqueduct on the Ohio and Erie Canal built near Akron, Ohio in 1859. This structure was sized to carry a trough of water 22' wide and 4' deep.

He moved his bridge manufacturing plant to the Boston area early in the Civil War and built hundreds of bridges throughout the northeast. The Moseley iron arch bridge, Fig. 2, built in Lawrence, Massachusetts across the North canal in 1864 to service the Upper Pacific Mills was one of eight Moseley's built across the north and south canals of the city for the Essex Company. It is the oldest extant iron bridge in the Commonwealth of Massachusetts and one of the oldest riveted wrought iron bridges in the United States. It is one of four bridges built by Moseley still in existence.

Figure 2 1864 Moseley across the North Canal

The bridge served its function well until the 1920s when a wooden structure was placed in the canal to support the iron structure. It is not known what condition the bridge was in at this time but from that time until 1989 the bridge arches did not carry any load. The arches served only to separate the walkways and the roadway. In the 1980s the bridge was only used from pedestrian traffic. The recording team of the Historic American Engineering Record (HAER) visited the bridge in 1976 and documented the bridge.

Moseley's iron arch evolved over time from the arch shown in his patent application. In 1859, "a radical improvement was made in the bridge, greatly increasing its strength and stiffness."(Moseley 1863) This was the addition of the counterarches, which are the most distinctive features of most Moseley's built after the Civil War. He wrote in 1867 that:

> In no other form can iron be arranged to bear a greater burden than in this (his arches). By actual tests, every inch of iron in the cross section of the Arches will bear 15,000 lbs; but in practice we provide for a weight of not less than four times the required burden, calculated, moreover, at a pressure of 7,000 lbs. per square inch of section.
>
> ...in Highway Bridges, the cross-section of the chords has one-third the number of inches in that of the Arches; and in Railroad Bridges, one-half the number of inches in cross-section. The tensile strength of an inch of iron averages 60,000 lbs.: but, as will be seen, our calculations are based upon a tensile strain of 21,000 lbs. per square inch (in Railroad Bridges 14,000 lbs.), for four times the actual burthen (Burden).

He was obviously using a high safety factor by placing a load on his bridge four times that expected and still keeping the working stress to 21,000 psi. It appears from the record that most of Moseley's bridges did not fail from a lack of material in the arches but in a want for lateral stability which is common in pony truss type bridge structures. Frequently, Moseley would place outriggers supported from extensions of the floor beams with struts to the top chords. When his bridges were used with sidewalks, however, he could not place the outriggers and had to resort to the system used at the subject bridge. As shown in Figure 3 he effectively cantilevered his lateral support system off a strengthened floor beam.

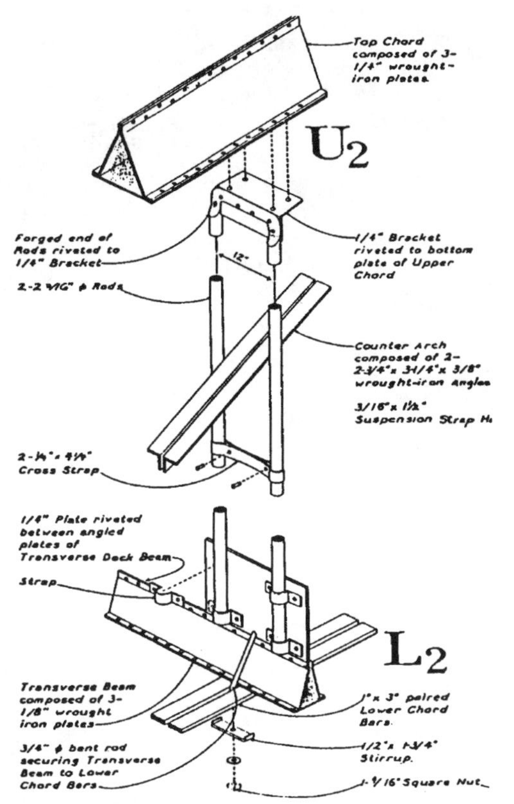

Figure 3 Cantilevered Lateral Support System (HAER)

Collapse in 1989

The wooden understructure collapsed in the summer of 1989 and with the loss in support the southerly arch buckled and the suspender straps pulled out of the top chord member. The owner of the bridge was in the process of scrapping it when I became aware of the collapse. I asked if he could have the bridge transported to Merrimack College where the students and I would like to restore it for use as a footbridge in the area. The owners were agreeable and on the following Saturday the bridge parts were trucked to the campus.

Figure 4 Collapsed Bridge

Figure 5 Bridge when it arrived at the Merrimack College

Rehabilitation

The first step in the rehabilitation was to spread out the parts to determine what was usable with some repair. Upon a detailed examination we decided that most of the top chord plates were reusable except for the ends and places where the contractor cut the arches in half with an acetylene torch and the point where the arch had buckled. We used some sections from the ends and spliced them into the middle sections. This required replacing the end sections with steel.

All plates connecting the straps to the top chords were replaced with steel, as were 30 of 94 of the suspender straps. The lower tension chords all necked down at locations where iron breams were strapped to them to support the cantilevered lateral support system (Figure 3). We welded 2" x 1/2" straps continuously along the bottom of the straps to ensure adequate section. The counterarches were badly bent during the collapse and shipping to Merrimack. The entire deck structure had to be replaced as all the wood and iron members were beyond repair.

Figure 6 Pieces Spread out for Observation

Figure 7 New Steel End Sections

Most of the rehabiliation was accomplished in the fall of 1990 and spring of 1991. The arches were erect by November 1990 as shown in Figure 8.

Figure 8 Arches Erect and Carrying their own Dead Weight

The connections were firmed up in the spring and the deck applied in the summer of 1991. A new lateral support system was required as the original materials which Moseley used to brace his bridge were damaged beyond repair. We decided to use two tubular steel sections welded to brackets, which had in turn been welded to the arch sections at the third points. This system worked well and also served as an ideal place to mount the original nameplates.

Figure 9 Bridge with Deck and Lateral Support Members

All work on the trusses had been accomplished with contributed labor, materials and equipment. We began discussions with the city of Lawrence to bring the bridge back to its original home. Unfortunately some of the owners would not accept it, unless it was able to support emergency vehicles. After several years of trying to gain support of the city and adjacent landowners, we decided to approach the college to find a home for the bridge on campus.

The original master plan for the college called for an arched bridge to span a reflecting pond in the middle of campus. We used this concept and obtained approval of the Board of Trustees to place the restored Moseley on campus, provided it did not cost the college any money. In the spring of 1995 we placed the foundations, moved the bridge onto the foundations and then constructed a reflecting pond under the bridge, Figures10, 11.

Figure 10 Moving the Bridge

Figure 11 The Bridge in Place

Water filled the pond in August 1995, and the site was landscaped shortly thereafter. Two contributed fountains were added in September and a commemorative plaque was mounted on a contributed granite block. The bridge was dedicated and blessed in early December and now serves the students, faculty and friends of Merrimack College. It is fast becoming an icon of the college and is used in many publications of the college as an example of the bond that exists between the past and the future.

The bridge was nominated as a National Historic Civil Engineering Landmark in the fall of 1997 and approved by the Board in the spring of 1998. A dedication ceremony is being scheduled for the fall of 1998. With this designation the Moseley joins the Dunlaps Creek Cast-Iron Bridge 1839, The Whipple Bridge 1855 (Union College), The Bollman Truss (Savage, MD), The Fink Truss 1852 (Lynchburg, VA), and The Bridges of Keeseville, NY (1874 Pratt Truss) as surviving examples of mid 19th Century Cast and Wrought-iron bridges.

References:

Moseley, Robert B., Moseley Family (A Genealogy), 1948.
Moseley, Thomas W. H., Iron, New Enterprise in its Manufacture and Applications to Building, Boston, 1863.
Moseley, Thomas W. H., Iron Bridges-Roofs &c Manufactured by Moseley Iron Building Works, 1867.
Railroad Record, Cincinnati, Ohio, March 26, 1857.

The Covered Bridges of Southwestern New Hampshire

By

Francis E. Griggs, Jr.[1]

Abstract

New Hampshire is second only to Vermont in the number of covered bridges still in use. Of the more than 50 covered bridges almost half of them (24) are located in the southwestern portion of the state bounded by I-93 on the East, I-89 on the north, the Connecticut River on the West and the Massachusetts State line on the South. In the late 1980s and 1990s a great deal of work has taken place in this region to preserve this important part of the heritage of 19th century New Hampshire. This paper describes these bridges and the men who designed and built them and gives a brief history of wooden truss building in the United States.

The Bridge Builders

Wooden truss bridge building in the United States dates from the late 18th century. The first major wooden truss bridge builder was Timothy Palmer, a Massachusetts millwright and sometimes architect. His first bridge of any significance, the Essex-Merrimack Bridge, was near Newburyport, Massachusetts over the Merrimack River. A small island at this location divided the river and Palmer built an arched structure after a plan of Palladio, the 16th century architect from the southerly shore to the island. Another shorter span ran from the island to the north shore and contained a small lift span to permit the passage of masts of sailing ships. This bridge was not covered when it was built in 1792 but was covered later around 1810. Palmer went on to build other bridges across the Merrimack River at Lawrence and Haverhill just upstream from his first bridge as well as ones over the Connecticut and Keenebec Rivers. He was issued a patent on his bridge in 1797. The bridge Palmer is best known for is the Permanent Bridge over the Schuylkill River

[1] Director Quality Assurance, Cough, Harbour & Associates LLP, III Winners Circle, Albany, NY 12205

near Philadelphia, Pennsylvania. This 550 foot long, three span bridge, was completed in

Figure 1 Palmer's First Long Span Wooden Bridge

1805. With prodding from the owners of the bridge, Palmer reluctantly agreed to cover it with a roof and siding. From that time on most long span and many shorter span wooden bridges were covered.

Theodore Burr was the next major wooden bridge builder. He added an arch to his truss patterns and this permitted him to build very long spans. He built many of these bridges around the country including a wooden suspension bridge at Schenectady, New York over the Mohawk River and the Union Bridge over the Hudson River at Troy, New York. The Union Bridge, the first bridge to cross the Hudson River between New York City and Troy, NY was in service over 100 years until it was destroyed by fire in the early 1900s.

Ithiel Town, the inventor of the Town Truss, was like Palmer an architect and a master of wood construction in buildings and churches. He patented his Lattice Truss in 1820. It was easy to build and used similarly sized elements for all its diagonal members. The latticework was connected with wooden pegs, call treenails. A large proportion of wooden bridges that remain are of the Town lattice pattern. His bridges were all covered and later the trusses were doubled and used for railroad bridges.

Stephen H. Long followed Town. He was a Hopkinton, New Hampshire native, who after graduating from Dartmouth College, entered the United States Army, n. He received several patents on wooden bridges between the years of 1830 and 1858. He was the first wooden bridge builder to use analytical means to size the members of this truss. His brother Moses, and his cousins the Childs brothers, handled the business end of his bridge building enterprise. They either built bridges to Long's patent or collected royalties from others using his patent.

Figure 2 Long, Howe, Whipple

William Howe, born in Winchester, Massachusetts, patented a truss pattern in 1840 that for the first time used wrought iron as a major load-carrying element. Iron rods replaced the wooden verticals, which are in tension, for the first time gave the bridge builder the opportunity to effectively adjust the camber, the upward arching of the deck. Prior to this attempts were made to adjust camber by placing wooden wedges at select junction points of the truss.

Squire Whipple, born in Hartwick, Massachusetts, while known better for his iron bridges, also designed and built several long span covered bridges in the 1850s. Two of these were across the Mohawk River in Schenectady, New York and one was in Trenton Falls, New York. He used his own double intersection truss pattern and his trusses were the first ones to be designed using scientific methods.

Other pioneer wood bridge builders were Lewis Wernwag who built the "Colossus" over the Schuylkill River with a span of over 340 feet in 1812 and Nicholas Powers from Vermont who built the Blenheim, New York bridge over Schoharie Creek with a span of over 215 feet in 1855.

Long and Town were both entrepreneurs who used agents to sell the rights to their patents throughout the eastern part of the country. Peter Paddleford of Littleton, NH used Long's patent. He later developed his own truss pattern called the Paddleford Truss and built many of them throughout the state as well as in Vermont. His was a simple variation of Long's pattern whereby the upper end of his tension diagonals did not end at a panel point. The truss was built both with and without supplementary arches. There are no Paddleford Trusses in the southwestern part of New Hampshire.

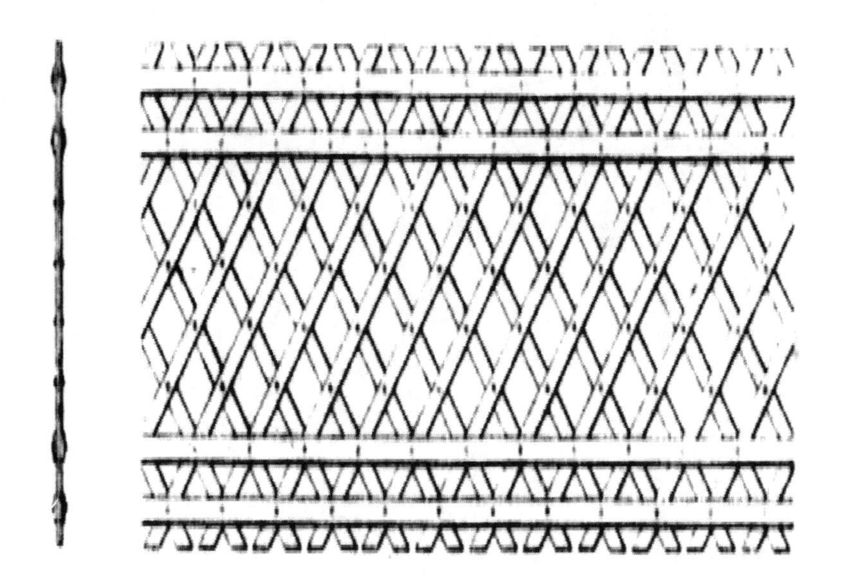

Figure 3 Town Truss

The major bridging challenge in the area was the Connecticut River, which separates Vermont from New Hampshire. The first wooden bridge recorded in the country was Enoch Hale's bridge connecting Walpole, NH with Bellows Falls, VT. The river narrowed down there passing through a gorge at this location with an island in the middle. Hale placed a pier on the island and built a two span beam bridge with supplementary struts rising from the abutments and the center pier to support the beams.

The Present

With this brief background I will now describe the 24 bridges which are the topic of this paper. The Covered Bridges still in use in the southwestern part of the state are located on Figure 4 and described in Table 1.

Starting in the southwestern corner of the state with bridges over the Ashuelot and/or its branches we have the Ashuelot, Coombs, Slate (burned in 1993 with plans underway to rebuild it), Thompsons (rebuilt 1993), Sawyers Crossing (rebuilt in 1991), Carleton (rebuilt in 1997). All of these with the exception of the Carleton are

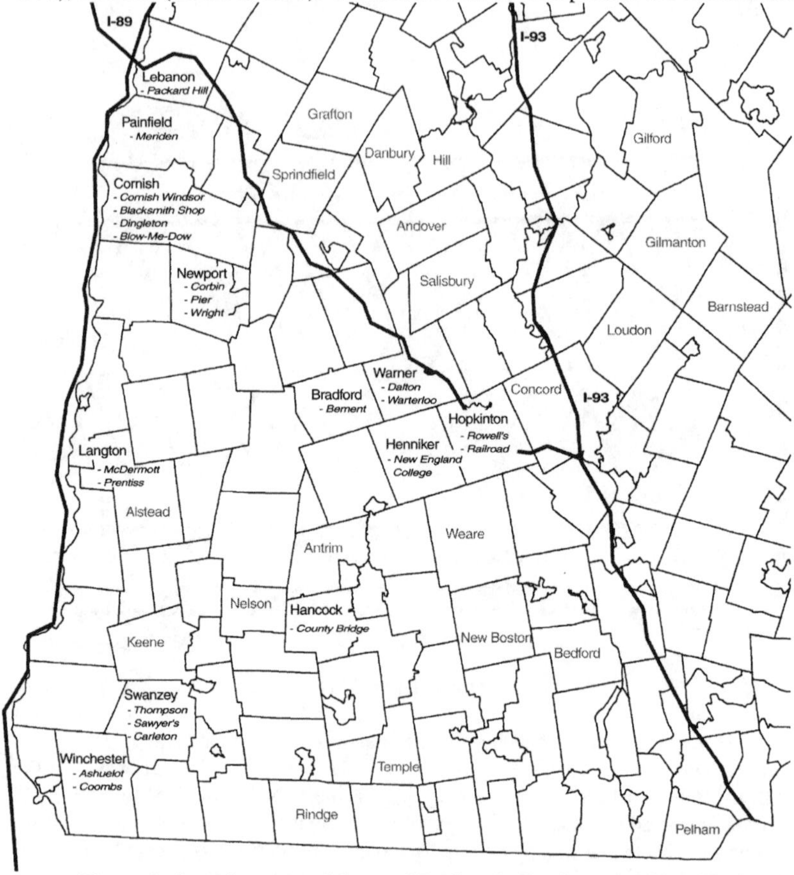

Figure 4 Location map of Covered Bridges in Southwestern New Hampshire

Town Trusses. The Carleton is a 69' span made of Hemlock. Its portal, according to local tradition, was sized to pass a wagonload of hay. It used iron verticals and open stonework abutments. I had the pleasure of being the engineer on the restoration of this bridge in 1997.

Moving north we have the McDermott and Prentiss Bridges in the towns of Langdon and Ackworth. The Prentiss Bridge is the shortest span covered bridge in the state and is a modified Town pattern. The McDermott is a Town Lattice with an assisting arch. It is the fourth bridge to stand at this location with previous bridges dating 1790, 1814, and 1840. This bridge, built in 1869, is closed to vehicular traffic.

The Town of Newport is challenging the Town of Swanzey as the covered bridge capital of New Hampshire. The town boasts the Pier, Wrights and Corbin Bridges. The Corbin Bridge was destroyed by fire in 1993 but was completely rebuilt

Date NR listed	Bridge Name	River Crossing	Nearest Town	Nearest Street/H'way	Total span	date built
11/14/74	Sawyers Crossing	Ashuelot	Swanzey	off NH 32	158' 5"	1859
6/10/75	Carleton	Ashuelot	E. Swanzey	Carleton Road	67' 3"	1869
6/10/75	Pier RR Bridge	Sugar	Newport	Chandler Road	216' 7"	1907
6/10/75	Wright's	Sugar	Claremont	Chandler Road	123' 9"	1906
11/21/76	Bement	Warner	Bradford	Center Road	60'	1854
11/21/76	Dalton	Warner	Warner	Joppa Road	76' 6"	1853
11/21/76	Rowell's	Contoocook	W.Hopkinton	Clement Hill Road	164' 6"	1853
11/21/76	Coombs	Ashuelot	Winchester	off NH10	106' 6"	1837
11/21/76	Waterloo	Warner	Waterloo	Newmarket Road	76' 4"	1840
11/21/76	Cornish Windsor	Connecticut	Cornish	West NH12A	449' 5"	1866
5/19/78	Blow-Me-Down	Blow me Down	Plainfield	off NH12A	85' 9"	1877
5/22/78	Blacksmith Shop	Mill Brook	Cornish City	off NH12A	91'	1881
11/8/78	Dingleton	Mill Brook	Cornish Mills	off NH12A	77' 9"	1882
11/14/78	Slate (**fire 1993**)	Ashuelot	Westport	off NH10	142' 3"	1862
1/11/80	Contoocok RR Bridge	Contoocook	Hopkinton	Off NH 103	140' 1"	1849
2/29/80	Thompson	Ashuelot	Swanzey	Main Street	136' 10"	1832
8/27/80	Meriden	Blood Brook	Meriden	Colby Hill Road	80'	1880
2/20/81	Ashuelot	Ashuelot	Winchester	NH119	169'	1864
	Packard Hill	Mascoma	Lebanon	Riverside Drive	76'	**1991**
	Prentiss	Great Brook	Langdon	South NH12A	34' 6	1874
	McDermott	Cold	Langdon	North NH12A	81'	1869
	Corbin (**fire 1993**)	Sugar	Newport	West of NH10	96' 2"	**1994**
	Henniker	Contoocook	Henniker	South of NH9	136' 7"	**1972**
	County	Contoocook	Greenfield	East US202	86' 6"	**1937**

Table 1 Covered Bridges of Southwestern New Hampshire

in 1994. The Pier is a double Town lattice (twin lattices both sides) and Wright's is an arch assisted Town Lattice. Both of these bridges were built for railroad purposes and are currently used for pedestrian traffic. The town is planning the construction of another covered bridge utilizing a typical 19th century pattern.

The Cornish area boasts the Blacksmith Shop, Dingleton, Blow-me Down and Cornish Windsor Bridges. The first three bridges are all less than 100' spans crossing small brooks and are multiple kingpost (Pratt) trusses. Milton Graton repaired the first two in 1983. The Cornish Windsor Bridge, in two spans, is the longest (449' 5") covered bridge in the United States and is a modified Town Lattice Truss.

Just north of Cornish the Meriden and Packard Hill Bridges are found. The Meriden is a multiple Kingpost (Pratt) truss and is the third bridge to exist at this site. It is now supported with steel beams so that the wood trusses carry very little of the load. The Packard Hill Bridge is a re-creation by the sons of Milton Graton. They built the new bridge to the same Howe Truss pattern as previously existed using similar tools and techniques of the 19th century.

Following I-89 to the southeast we find the Waterloo, Bement, and Dalton Bridges crossing the Warner River near Warner. The longest span of the three is only 65'. The Waterloo is a modified Town Truss; the Bement is a Long Truss, and the Dalton is a Queenpost truss superimposed on a Long Truss. It is likely that the Queenpost Truss was added to the Long Truss at a later date. Originally constructed in 1853, it is one of the oldest wooden trusses in the state

Hopkinton, the home of Stephen H. Long, houses the Rowell's and Railroad Bridge. Rowell's is Long truss with assisting wooden arches. The bridge was also built around 1853 making it a very old structure. The Railroad Bridge was originally built to carry the Concord and Claremont Railroad over the Contoocook River in Hopkinton. It is a twin Town Truss and like some other railroad bridges has a flat roof. It was originally built in 1849 so it claims the title of the oldest wooden bridge in the State. It is currently under the jurisdiction of the NH Division of Historic Resources.

Figure 5 Carleton Bridge, Swanzey, NH

The last bridge in our tour through southwestern New Hampshire is the County Bridge located in Greenfield and crossing the Contoocook River. It is a modern wooden bridge dating from 1937 and utilizes a Howe Truss. It uses modern timber framing connections and gusset plates of steel. It is the youngest bridge in our inventory and has the look of a 20th century bridge.

Figure 6 Cornish Windsor Bridge

The 24 bridges within the boundaries of this inventory range from those in span from 34' 6" (Prentiss) to 449' 5" (Cornish Windsor). They were built for railroad, major roadways and minor roadways. They used the truss patterns of the 19th century: the Town, Long, Howe, Pratt and arch supported variations of these styles. The good news is that many of these bridges have been restored or recreated within the last two decades. Prior to the 1966 Historic Preservation Act bridges were disappearing at an alarming rate. As seen in Table 1 many of the bridges are on the National Register of Historic Places and receive the protection associated with that designation.

Perhaps the greatest force in the effort to preserve and restore the covered bridges of the state came from Milton Graton of New Ashland, NH. Starting in the 1950s he, and his sons, restored many bridges in NH including the Lydonville (1959), Stoughton (1963), Groveton (1965), Bartlett (1966), Durgin (1966), Turkey Jim (1970s), Henniker (1972), Blair (1977) and the Bedell (1978). His son, Arnold Graton, has now taken over the business and continues in the path that his father forged. In southwestern New Hampshire he built the Packard Bridge and the Corbin Bridge. Other covered bridge restoration companies, such as the Wright Construction Company and Chesterfield Associates, have trained skilled craftsmen with an appreciation of the materials and methods of the past. They, along with Arnold

Graton carry on the tradition of Milton Graton and the builders of the previous century.

Conclusion

It is clear that the small towns throughout New Hampshire are taking greater interest in their covered bridges. The selectmen of Swanzey and Newport are especially to be commended for their commitment to the restoration of their covered bridges. The investment the state continues to make in bridge restoration is also to be commended. The Cornish Windsor Bridge, with its glulam lower chords, still stands as a testament to the ability of James Tasker and Bela Fletcher, its original builders. In 1866, these men, even though illiterate, but having the intuition and knowledge of wood as a material, initiated a project which even today would challenge the best engineers using the best computers.

The continued preservation of these icons of early American Engineering should be promoted so that our children, and our children's children, can appreciate the ingenuity of the builders of 19th century America.

THE LOWELL NATIONAL HISTORIC PARK
BIRTHPLACE OF THE AMERICAN INDUSTRIAL REVOLUTION

Lawrence Richard Keats, P.E., Member, ASCE[1]

Abstract

The 1998 Convention of the American Society of Civil Engineers in Boston presents an opportunity for attendees to visit the Lowell National Historic Park and observe some of the facilities that marked the beginning of the industrial revolution in America. The park was created to interpret the history of the industrial revolution in Lowell, Massachusetts through the preservation of historic cotton mills, canals, gatehouses and other facilities related to the manufacture of textiles through the use of waterpower. This visit to the park has been organized by the History and Heritage Committee of the Boston Society of Civil Engineers Section, American Society of Civil Engineers.

Narrative

The City of Lowell is situated at the confluence of the Merrimack and Concord Rivers, 26 miles northwest of Boston. Here, the Merrimack River, flowing south out of New Hampshire, descends over Pawtucket Falls approximately 32 feet before flowing east and emptying into the Atlantic Ocean at Newburyport, Massachusetts. Waterpower for cloth weaving mills was harnessed through a system of dams and gates at the falls, and approximately 5 ½ miles of canals that brought water to various mill buildings. Similar mill complexes were later established up and down the Merrimack River Valley in Nashua and Manchester, New Hampshire and Lawrence and Haverhill, Massachusetts, as well as others sites throughout New England and northeast America.

Patrick Tracy Jackson and Nathan Appleton, Boston mercantilists, began the development of the mill complex at Lowell in 1822. The surrounding village was established as a town in 1826, and became a city in 1836. It was named after

[1]President, Lawrence R. Keats Associates, Inc., Civil Engineers 28 Church Street, Suite #10, Winchester, Massachusetts 01890

Francis Cabot Lowell, Jackson's brother-in-law and his partner in the development of an earlier mill on the Charles River in Waltham, Massachusetts.

Francis Cabot Lowell, who was born in Newburyport in 1775, graduated from Harvard University in 1793 and became a merchant in Boston. In 1810 he visited England and after observing the manufacture of cotton cloth there, wished to establish similar manufacturing facilities in America. He was hampered by the reluctance of the English to share their technological advantage, and later by the War of 1812 from importing any machinery. Despite these obstacles, Lowell designed and built a power loom based on his memory of machinery that he had observed in England. He partnered with his brother-in-law, Jackson, and together they established their first mill, the Boston Manufacturing Company at Waltham in 1814. It is reported that this new mill was the first in America to contain all the processes for the manufacture of finish cloth in one plant. The new mill was located just 7 miles west of Boston, where raw materials and finished goods could be easily transported to the mercantile hub of Boston by way of the Charles River.

Following Lowell's death in 1817, Jackson, recognizing the need for a new site with greater power potential for the growing business, acquired land around Pawtucket Falls in the village of East Chelmsford, or what is now the City of Lowell. Although the new mill site lacked a direct river connection to Boston, it was connected artificially by water through the Middlesex Canal, one of the earliest canals constructed in the United States and by the Middlesex Turnpike, a private turnpike. Later, in 1830 Jackson obtained a charter for the construction of a new railroad from Boston to Lowell.

The beginning of the nineteenth century was a period when private development of canals and turnpikes was encouraged by the federal and state governments in order to promote trade between mercantile centers. At the time, although the need was recognized, the governments did not have the resources to develop similar public transportation facilities. The Middlesex Canal and the Middlesex Turnpike were chartered and constructed as part of a larger system of canals and roads intended to develop trade to and from Boston and the upper Merrimack River Valley. The later development of the Boston and Lowell Railroad, plus other railroads with their greater efficiency for moving goods and people, resulted in the eventual demise of the canal and turnpike companies. Remnants of the canal and northern portions of the turnpike remain and can still be visited. One portion of the Middlesex Canal, the stone pier and abutments of an aqueduct over the Shawsheen River was declared a National Historic Civil Engineering Landmark in August, 1967 by the American Society of Civil Engineers.

The Lowell National Historic Park was established in 1978. Some of the facilities preserved and maintained at the park include cotton mills, gate houses, canals, worker housing and water power exhibits. Some of the major facilities that are maintained to illustrate the story of the industrial revolution in Lowell include:

- Guard Locks, located at the entrance to the Pawtucket Canal from the Merrimack River. The Pawtucket Canal was the first canal of the Lowell Mill complex. Here is located a floodgate, constructed in 1850 to protect the city from high water in the river. So-called "Francis' Folly" after its designer, James B. Francis, Chief Engineer for the Lowell Machine Shop, the floodgate was twice used in 1852 and 1936 to protect the city from flooding.

- Lower Locks, located at the outlet of the Pawtucket Canal at the Concord River.

- Pawtucket Gatehouse, constructed between 1846 – 48 and still in use today. It is used to control water flow into a then, new Northern Canal. The development of this canal system allowed a 50% increase of waterpower potential, and construction of additional mill buildings.

- The Suffolk Mill Turbine Exhibit located in the former Suffolk Manufacturing Company building. The exhibit contains nineteenth century water turbines, including one maintained in operating condition.

- The Boott Cotton Mills Museum containing operating power looms in an 1920s-era factory weave room. The museum also contains exhibits about the mill girls, immigrants and others who made up the work force at the mills.

Works cited

Americana Corporation, *The Encyclopedia Americana*, New York, Americana, 1962.

National Park Service, *Handbook 140, Lowell, The Story of an Industrial City, A Guide to Lowell National Historic Park and Lowell Heritage State Park*, National Park Service, U. S. Department of the Interior, Washington, D.C., 1996.

Raphael, Thomas, *Of Canals, Turnpikes and Railroads,* Unpublished narrative.

Seaburg, Carl, Alan Seaburg, and Thomas Dunhill, *The Incredible Ditch, A Bicentennial History of the Middlesex Canal*, Medford, Miniver, 1997.

150 Years of Underground Design and Construction

S. Trent Parkhill[1]

During the past 150 years, America progressed from a collection of rural farm-based communities to an urban industrialized society. Rapid population growth and the concentration of people in urban areas created demands for expanded transportation systems, taller buildings and construction over poorer subsurface conditions. These demands helped to transform underground design and construction from an art practiced by contractors and architects to a profession practiced by geotechnical engineers and geotechnical specialty contractors.

1848-1879: In 1848, both America and the civil engineering profession were experiencing major changes. The addition of new territory and the discovery of gold created a wave of westward migration and demand for improved transportation routes to the west. These demands would eventually lead to major excavation projects including the Panama Canal and the Transcontinental Railroad. Against this backdrop, on July 3, 1848, the Boston Society of Civil Engineers (BSCE) held its first meeting, creating the first engineering society in the United States. BSCE, and the other engineering societies that followed, provided a forum for engineers to exchange ideas and experience, and helped to establish a common vision for the development of the new profession. In 1848, building foundation design and construction had advanced little past that of the fifteenth century. However, by 1879, several important advances had occurred including the introduction of the "allowable bearing pressure" concept and the development of the first concrete and steel spread footings. While some advances were quickly adopted, many were not. For example, although the steam pile hammer was developed in 1845, most of the piles driven between 1848 and 1879 were driven with a steam-raised drop hammer and some pile rigs were still raising the hammer with animal or human power.

---

[1] Senior Engineer, Haley & Aldrich, Inc., 465 Medford Street- Suite 2200, Boston, MA 02129-1400

1880-1924: The period from 1880 to 1924 was a time of rapid population growth and economic expansion, with steel replacing cast-iron and wrought-iron, and electric and gasoline power beginning to replace the steam engine. Rapid increases in land costs in many American cities, combined with the availability of steel and the invention of the elevator, produced the world's first skyscrapers. As building heights increased, the use of deep foundations also increased. With the introduction of concrete and mass-produced steel, more than a dozen new pile types were developed and the first hand-dug caissons were used to support buildings. By the end of this period, machine-drilled caissons had been used. While some engineers began to be concerned about the condition of the soils below foundation bearing levels, the lack of methods for obtaining undisturbed samples and testing their properties helped to prevent the development of a reliable quantitative approach to underground design. Foundation design practices continued to be based on experience, with practitioners attempting to draw rational conclusions and design approaches from these experiences.

1925-1945: In 1925, Karl Terzaghi published *Erdbaumechanik* (Soil Mechanics), the event marked by many as the birth of soil mechanics. By the end of the year, Terzaghi's concepts had been widely distributed in English, through the printing of several articles in Engineering News-Record. Despite the difficult economic times that followed in 1929, federal spending on infrastructure projects helped to support research by Terzaghi, Arthur Casagrande and others, and thus to rapidly expand the new field. This research demonstrated the changes in clay properties caused by disturbance and showed how the behavior of clay could be predicted. This research also led to the development of the split spoon sampler (1926) and the Shelby Tube Sampler, and the practice of taking intact soil samples in exploration programs. With Terzaghi's unifying concepts and terminology, and the ability to obtain undisturbed samples for testing, soil mechanics and foundation design research flourished during this period. Rational design could now begin to replace the pre-1925 empirical rules and observations. Technological advances of the period included the introduction of low-pressure pneumatic tires (1930) and significant improvements in caterpillar tractors. With these improvements, equipment could now move quickly around construction sites boosting productivity to a level that finally brought an end to the use of animal power and large scale hand labor in construction. Advances in science and equipment also spurred the development of soil improvement techniques, including vibroflotation, vertical sand drains, wick drains and rubber-tired rollers. Advances in deep foundations included the introduction of many new pile types, including the Raymond Step-Tapered Pile and the Franki Pile, and the development of the first crawler-mounted pile driving rigs.

1946-1960: As World War II ended, the conversion of America's massive wartime industrial base helped to create a period of rapid economic expansion. With this growth in prosperity and the pent-up demand from the depression and war, Americans rushed to buy cars. The road construction that followed prompted research into soil improvement and helped to stimulate the development of many consulting firms specializing in soil mechanics and foundation engineering. In 1948, the Second International Conference on Soil Mechanics and Foundation Engineering was held in Rotterdam, the Unified Soil Classification system was proposed by Arthur Casagrande and *Soil Mechanics in Engineering Practice* was published by Terzaghi. Soil mechanics had acquired considerable momentum, with both research and practice solidly established around the world. As taller buildings were erected and the number of cars increased, demand for parking space resulted in an increase in the typical number of basement levels in these buildings. By 1960, these changes had helped to stimulate developments in deep temporary lateral support systems and higher capacity piles and caissons.

1961-1998: The construction of the U.S. Interstate Highway system and the technology spinoffs from the space program stimulated the development of underground design and construction during this period. The dominant technological advance of this period was the development of the computer. The iterative calculations made possible by computers resulted in many advances in underground design, including the development of the pile wave equation, more rigorous slope stability analysis methods and the finite element method of analysis.

Summary: The evolution of underground engineering and construction has been shaped by the interaction between changes in demand, advances in science, and the development of new materials and sources of power. Demands placed on underground engineers by the growth in the construction of canals, railroads, roads and tall buildings have led to the creation of rational methods of design and a new branch of civil engineering.

Pioneers in Soil Mechanics: The Harvard/MIT Heritage

by

Anni H. Autio, P.E., M.ASCE[1] and Michael A. McCaffrey, P.E., M.ASCE[2]

Contributions by Boston's pioneers in the field of soil mechanics at Harvard University (Harvard) and the Massachusetts Institute of Technology (MIT) shaped and advanced the geotechnical engineering profession. As part of BSCES's 150th anniversary celebration, a history-in-the-making panel discussion will be held at the 1998 ASCE Convention to commemorate the work of Boston's pioneers in soil mechanics. The panel discussion will be moderated by Ronald C. Hirschfeld (1998 ASCE Honorary Member) and will include personal remarks by a distinguished panel of invited geotechnical engineers. Each panelist shares a connection with and will share memories of, the pioneers in soil mechanics from Harvard and MIT. Personal anecdotes from those whose careers were influenced by pioneers such as Karl Terzaghi, Arthur Casagrande, Donald W. Taylor, and T. William Lambe will be transcribed into an article to be published in a future issue of *Civil Engineering Practice: Journal of the Boston Society of Civil Engineers Section/ASCE*. Invited participants include: Harl P. Aldrich, Jr., Ralph E. Fadum, Walter Ferris, John A. Focht, Jr., Donald T. Goldberg, James P. Gould, Ronald C. Hirschfeld, Ralph B. Peck, Steve J. Poulos, William Shannon, and Robert V. Whitman. An open discussion with the audience will follow the panel presentation. Peer-reviewed discussion papers will be accepted after the conference from those who want to publish their remarks with the session proceedings.

Biographies of four highly distinguished civil engineers from Harvard and MIT are presented below. The lifetime contributions by Terzaghi, Casagrande, Taylor, and Lambe in the field of soil mechanics represent the beginning of the Harvard and MIT heritage that has unfolded new generations of geotechnical engineers throughout the world. More has yet to be written.

---

[1]Project Manager, Camp Dresser & McKee Inc., 10 Cambridge Center, Cambridge, MA 02142 (1997-1998 BSCES/ASCE President and 150th Anniversary Chair for BSCES/ASCE)

[2]Project Manager, GEI Consultants, Inc., 1021 Main Street, Winchester, MA 01890 (1997-1998 BSCES/ASCE Geotechnical Group Chair)

KARL TERZAGHI (1883-1963)[3]

Karl Terzaghi, founder of soil mechanics and a renowned civil engineer, was born in Austria in 1883. He received a conventional education at a military school in Hungary. His education was supplemented by his eager indulgence in such extra-curricular activities as the study of the geography of exotic lands and the invention of a system of trigonometry. His education was frequently punctuated by the perpetration of imaginative pranks; one of the last which came close to causing his expulsion from the Technical University of Graz. Owing largely to the intercession of one professor, Terzaghi was allowed to continue his studies, and in 1904 he received his engineering degree. Two years later, he returned to the University for an additional year of study, chiefly in geology.

Terzaghi began practicing civil engineering as a superintendent of construction in various parts of the old Austrian Empire and in czarist Russia. Many of his jobs involved earthwork and foundation engineering, at that time necessarily performed on the basis of local custom and empirical rules lacking general validity. With each new job, he recognized more clearly the need to replace the old "rules-of-thumb" with more rational approaches to the problems of earthwork and foundation engineering. With this objective, his rebellious and creative nature found a purpose that would occupy his thoughts and consume his energies until his death.

In January 1912, after receiving his doctorate from the Technical University in Graz, Terzaghi traveled to the United States in the hopes of obtaining insight into fundamental principles by means of a systematic study of the relation between foundation conditions. His plans were to use this insight in an evaluation of the performance of a large number of major structures, then recently completed, or still under construction, in this country. Terzaghi collected a wealth of interesting observational data during his visit, however, he failed to obtain the answers he sought before returning back to Austria in 1913.

Terzaghi was convinced of the inadequacy of case studies, supported only by theoretical insight, into the behavior of soils under conditions imposed by civil engineering operations. He embarked on a program of experimental and theoretical research designed to provide the required fundamental knowledge. Committed to his research after some interruptions, Terzaghi wrote this revealing letter to former Professor F. Wittenbauer on February 15, 1920:

---

[3]Written from a Memoir on Karl Terzaghi by Ruth D. Terzaghi, Research Fellow, Harvard University that was published July 1964 in the *Journal of the Boston Society of Civil Engineers*, Volume 51, No. 3.

> *At the beginning of March [1919], I listed on a single sheet of paper everything that we needed to know about the physical properties of clay in order to be in a position to treat the fundamentals of earthwork engineering on a scientific basis. My requirement seemed excessive even to myself, and I doubted that I would live to see all the questions answered. In April, I started to build primitive apparatus . . . and by the end of the month the first experiments were running. After six weeks of 12-hour days in which I experienced uninterrupted failures, my luck changed, and since that time I have not carried out a single experiment that did not provide the anticipated enlightenment. Today, in mid-February, I can say that I have solved all the problems of last March . . . It is really a remarkable dispensation of the fates that my old dream of a scientific basis of engineering geology has been fulfilled in a period of such general disaster, and in such an incredibly short time. I believe that the cares, misfortunes, and insecurity of my situation have multiplied my powers many times over.*

Five years later, the results of his work were presented to the engineering profession in a book, *Erdbaumechanik auf bodenphysikalischer Grundlage,* published in Vienna.

In 1925, Terzaghi received a long-awaited invitation to lecture at MIT. Over the next four years, Terzaghi continued his fundamental research while serving as a part-time consultant on a large number of important projects on which he was able to demonstrate the immediate practical value of his discoveries. His research in combination with his consulting activities gave rise to a steady stream of highly original and important papers, several which are published in the *Journal of the Boston Society of Civil Engineers.* In 1929 when Terzaghi left MIT to accept a professorship at the Technical University of Vienna, soil mechanics was established as an academic subject and had begun to revolutionize earthwork and foundation engineering.

Terzaghi returned to Cambridge in 1938 to accept a part-time lectureship at Harvard. The University appointed him Professor of the Practice of Civil Engineering in 1946. After his retirement in 1956, Terzaghi continued to lecture for several more years as an *emeritus* professor on engineering geology.

At Harvard, his consulting practice, which he limited to exceptionally difficult projects, spread to all parts of the globe, and continued undiminished until the end of 1960, when failing health confined him to his home. During the last three years of his life, Terzaghi wrote six important papers, a large number of illuminating discussions, and several reports on projects with which he had been intimately and actively connected during the preceding years.

Terzaghi received many honors during his long and active career. The first was BSCE's Clemens Herschel Award which he received in 1926 and again in 1942. He also received ASCE's Norman Medal on four occasions, in 1930, 1942, 1946, and 1955. He was a member of many professional associations and was elected honorary president of the International Society for Soil Mechanics and Foundation Engineering after having served as its president for 21 years.

ARTHUR CASAGRANDE (1902-1981)[4]

Born in Haidenschaft, Old Austria, in 1902, Arthur Casagrande received his Civil Engineering degree in 1924, and Doctor of Engineering degree in 1933, both from the Technical University in Vienna, Austria. From 1924 to 1926, he was an Assistant in Hydraulics at the Vienna Technical University; and from 1926 to 1932, he was a Research Assistant with the U.S. Bureau of Public Roads, assigned to the MIT where he assisted Professor Karl Terzaghi in his numerous research projects directed toward improving the apparatus and techniques for soil testing. In those years, Casagrande developed the liquid limit apparatus, the hydrometer test, the horizontal capillarity test, the consolidation apparatus, and the direct shear test equipment. Among his various research projects, he also conducted field investigations on frost action during a cooperative project of the U.S. Bureau of Public Roads and the New Hampshire State Highway Department. His criteria for frost susceptibility of soils, resulting from this project, were adopted by highway designers all over the world.

Casagrande took a brief leave to build a soil mechanics laboratory for Terzaghi in Vienna but returned to MIT in 1930 where he built his first triaxial apparatus. Between 1930 and 1932, Casagrande concentrated on research concerning shear strength and on consolidation tests on undisturbed clay. This resulted in his fundamental discovery that excess pore water pressures develop during shearing. In addition, he established a procedure for identifying the pre-consolidation pressure of clays, and of evaluating time curves by means of semi-logarithmic plots.

In 1932, Casagrande accepted a lectureship at Harvard University, where he taught a two-semester course in soil mechanics and a course in foundation engineering. The following year, he added a course in soil testing. He received his Doctor of Science degree from the Technical University in Vienna in 1933 and returned to Harvard in 1934 as an Assistant Professor. In 1935, he started a course on seepage and groundwater movement and also handled his first assignment for the U.S. Army Corps of Engineers--to investigate the safety against liquefaction of the fine sand in the foundation of the Franklin Falls Dam in New Hampshire, and to advise on the control of seepage through the foundation of this dam.

In 1940, Harvard promoted Casagrande to Associate Professor. Between 1942 and 1944, Casagrande trained about 400 officers of the U.S. Army Corps of Engineers, on the soil mechanics aspects of airfield construction. It was in connection with these training courses that

[4]Written from a Memoir on Arthur Casagrande by Leo Casagrande and Dirk R. Casagrande, that was published in the Spring 1982 issue of the *Journal of the Boston Society of Civil Engineers Section/ASCE*, Volume 68, No. 1.

Casagrande developed his soil classification system, which was later adopted by the Corps and by the Bureau of Reclamation, with some modifications, as the Unified Classification System.

In 1946, Casagrande became Gordon McKay Professor of Soil Mechanics and Foundation Engineering. In 1973, he retired from Harvard and became professor *emeritus*. In addition to his consulting activities, Casagrande remained active as an educator, lecturer and author. He wrote or co-authored more than 100 professional papers.

During his nearly five decades of consulting practice, he acted as consultant to numerous federal, state, municipal and private organizations on the foundations of a great variety of structures, on groundwater problems, and particularly on the design and construction of earth and rockfill dams and dikes. These projects took him to all parts of the world. Included among his many projects were: design of foundation for the Prudential Tower in Boston; a comprehensive study of the stability of the slopes of the Panama Canal and of possible sea level conversion of the canal; and design and construction of many of the world's largest hydropower dams, including the Oahe Dam in South Dakota, the Manicouagan 3 Dam in Canada, the Tarbela Dam in Pakistan, and the Itaipu Dam in Brazil. He was occasionally called upon to investigate the cause of foundation problems or failures, as e.g. the failure of the Teton Dam in Idaho.

Casagrande was elected as a Fellow or Honorary member to many professional societies. He served as President of BSCE and of the International Society for Soil Mechanics and Foundation Engineering. Casagrande received multiple honorary doctor's degrees and was the recipient of many awards and prizes, including the First Rankine Lecture (Institution of Civil Engineers, London), Terzaghi Award and Terzaghi Lecture (ASCE), Decoration for Distinguished Civilian Service (awarded by the Secretary of the Army), Award of Merit presented by the American Institute of Consulting Engineers, the Moles Award, and the Order of the Rio Branco of Brazil. In 1978, the U.S. Army Engineer Waterways Experiment Station in Vicksburg, Mississippi, dedicated to him a new geotechnical research facility.

DONALD WOOD TAYLOR (1900-1955)[5]

Donald Wood Taylor was born in Worcester, Massachusetts in 1900 and received his S.B. degree from Worcester Polytechnic Institute in 1922. Taylor worked nine years with the United States Coast and Geodetic Survey and with the New England Power Association before he joined the staff of the Civil Engineering Department at MIT in 1932 as a research assistant in soil mechanics in the Department of Civil and Sanitary Engineering. His promotion in 1934 to research associate was

[5]Written from Memoirs on D.W. Taylor published in *Soil Mechanics* (T. William Lambe and Robert V. Whitman, authors; John Wiley & Sons, publishers) 1969 and in the April 1956 issue of the *Journal of the Boston Society of Civil Engineers*, Volume 43, No. 2.

followed by successive appointments as assistant professor of soil mechanics in 1938 and , in 1944, as associate professor of soil mechanics. He served as head of the Soil Mechanics Division from 1938 until his death in 1955. In 1942, MIT awarded Taylor with a Master of Science degree.

Taylor was well known as a consultant in foundation engineering and worked with private firms such as The Southern California Edison Company and the New England Electric System. He also served on numerous consulting boards for the U.S. Army Corps of Engineers through which his consulting activities included work on flexible pavements and on three classified projects associated with atomic blast programs.

Taylor, a quiet and unassuming man, was well respected among his peers for his very careful and accurate research work. He made major contributions to the fundamentals of soil mechanics, especially on the topics of consolidation and shear strength of cohesive soils, and the stability of earth slopes. He authored numerous reports and technical papers and his 1948 textbook, *Fundamentals of Soil Mechanics,* is still widely used.

Taylor was active in both ASCE and BSCE, where he served as Chair of the Committee on Subsoils of Boston for many years. Just prior to his death in 1955, Taylor had been nominated for the Presidency of the BSCE. Between 1948 and 1953 Taylor served as the International Secretary for the International Society of Soil Mechanics and Foundation Engineering.

Throughout his career, Taylor's work, whether in teaching, research, or engineering, was characterized by the highest professional standards; by great care and attention to accuracy; and by an allegiance to the scientific method in his quest for new knowledge. His high standards were as evident in his personal life as they were in his professional career. A man of utmost integrity, Taylor was guided by a desire to meet problems and to treat people with fairness and full cooperation. His high character and scholarly attainments won him undeniable recognition in the field of soil mechanics, where he stood among the leaders of the century.

T. WILLIAM LAMBE (1920-)[6]

T. William "Bill" Lambe was born in Raleigh, North Carolina in 1920 and received his undergraduate education at North Carolina State University (B.S. 1942). He pursued graduate studies at MIT (MS 1944, Ph.D. 1948).

Between 1943 and 1945, Lambe had various employment including positions at Standard Oil of California and at Dames & Moore. In 1945, he was appointed to the academic staff of MIT where he remained until his retirement in 1981. Upon the untimely death of D.W. Taylor in 1955, Lambe became head of the Soil Mechanics Division and held that position, with only a few

[6]Biographical Sketch in the MIT Symposium Proceedings entitled *Past, Present and Future of Geotechnical Engineering* held September 24-25, 1981.

brief interruptions, until 1974. Early in this time, Lambe arranged for Karl Terzaghi to become a regular lecturer at MIT and for Laurits Bjerrum to spend extended periods of time at the Institute. In 1969, Lambe was named Edmund K. Turner Professor of Civil Engineering and occupied that chair until his retirement in 1981.

Lambe has a distinguished career as a teacher, researcher, lecturer and consultant. His earliest research involved capillary phenomena in soils. In the late 1940's and early 1950's, he headed a large research effort at MIT concerning the stabilization of soils by chemicals and other additives. This project carried Lambe to the study of the fundamental nature of clay particles and of collections of such particles. This research led not only to development of more effective and economical means for chemical stabilization of soils, but also to the practical use of clay liners for reservoirs storing chemical wastes and oil and to a much improved understanding of shear strength mechanisms.

With time, Lambe's interests turned more and more toward applied engineering. He became a consultant to many companies and on many continents -- in Jamaica, Iraq, Venezuela, Turkey, the Netherlands, Libya, Italy, Japan and Portugal. He became an active and effective advocate for the "stress path method" for approximate analysis of stability and deformation. Lambe also focused attention upon "prediction" as central to the role of the engineer, and arranged a series of important symposia during which experts from around the world were invited to predict -- in advance -- the outcome of some full-scale experiment in the field.

Lambe is well-known as the author of two popular textbooks: *Soil Testing*, first published in 1951, and *Soil Mechanics* (co-authored with Professor R.V. Whitman). Lambe is a lecturer par excellence, and has been invited as keynote speaker for Pan American, Australia-New Zealand, Asian, South East Asian and Brazilian Conferences. In 1973 in Moscow, he was General Reporter for one of the sessions of the 8th International Conference on Soil Mechanics and Foundation Engineering.

Many awards and honors have been bestowed upon Lambe throughout his career. Of these, in 1970, he was invited to give the Terzaghi Lecture to ASCE and in 1973, the Institution of Civil Engineers in London asked him to be the Rankine Lecturer. He has been elected to membership in the National Academy of Engineering.

ACKNOWLEDGEMENTS: This session is co-sponsored by GeoCongress '98. Special appreciation is extended to GeoCongress '98 Chair: Dr. Thomas Liu and to Drs. Harl Aldrich, Jr., Charles Ladd, and Ronald Hirschfeld for their guidance during planning and preparation of this session.

REFERENCES

1. *Casagrande, Arthur* 1902-1981. Memoir Leo Casagrande and Dirk R. Casagrande. *Journal of the Boston Society of Civil Engineers Section/ASCE*, Volume 68, No. 1., Spring 1982. pp. 99-102.

2. *Lambe, T. William* 1920-. Biographical Sketch in the MIT Symposium Proceedings entitled *Past, Present and Future of Geotechnical Engineering* held September 24-25, 1981. Published March 1982. pp. i-ii.

3. Lambe, T. William and Whitman, Robert V. *Soil Mechanics*. Biographical Sketch on D. W. Taylor. John Wiley & Sons, 1969.

4. *Taylor, Donald Wood* 1900-1955. Memoir. *Journal of the Boston Society of Civil Engineers*, Volume 43, No. 2, April 1956. pp. 142-143.

5. *Terzaghi, Karl* . Memoir by Ruth D. Terzaghi, Research Fellow, Harvard University. *Journal of the Boston Society of Civil Engineers*, Volume 51, No. 3, July 1964. pp. 288-293.

SUMMARY OF
"THESE OLDE PYRAMIDS" PRESENTATION

R.J. Kachinsky[1], Dr. M. Lehner[2], Dr. Z. Hawass[3]

Background

The discussions will encompass activities by civil engineers which aid Egyptology. They are based on actual experience over the past 25 years in and around the site of the Gizeh pyramids and Sphinx. A great deal of collaboration and synergy between the totally different disciplines of Egyptology and Civil Engineering have been brought to bear on activities in this area with very successful outcomes.

This collaboration is continuing and will do so for as far into the future as one can see.

The principal participants in this presentation come from widely different backgrounds and perspectives. All have considerable knowledge in their fields and, more importantly, in-depth understanding and appreciation of the others.

Robert J. Kachinsky (F.-ASCE) is a Civil Engineer with a lifelong avocation in Egyptology. Dr.'s Lehner and Hawass are both world renown Egyptologists with a specialty in the monuments of the Gizeh Plateau. Both highly value the need for and methodologies of the specific aspects of Civil Engineering which benefit their work in gaining knowledge and protecting the magnificent monuments at Gizeh. This resulted in R.J. Kachinsky becoming associated with Dr. Hawass in his capacity of Project Manager

---

[1] R. J. Kachinsky, CDM International Inc., 10 Cambridge Center, Cambridge, MA 02142

[2] Dr. M. Lehner, Harvard University, c/o A.E.R.A., 16 Hudson St., Milton, MA 02186

[3] Dr. Z. Hawass, General Director of Giza Antiquities, Inspectorate of Giza Pyramids, Giza, Egypt

for the Cairo Wastewater Project. A relationship which expanded into numerous other collaborations, unrelated to the construction project, and which are still ongoing.

As a result of this collaboration, in the course of major construction of sewerage facilities in the Gizeh area, an enormous amount of significant archaeological information was obtained and properly recorded. Such valuable information would have been lost forever if the "old" approach of non-cooperation had been used and, significant delays in the construction would most likely have taken place. No delays were encountered! Therefore, the "new" approach of cooperation accomplished its intended goals. The key component was the establishment of clear understanding of each parties needs and objectives, and mutual confidence/trust.

Another project in which the author is involved includes water and wastewater work in Luxor, Egypt. Luxor is another major antiquities area of Egypt. A similar collaboration 4with the antiquities people there has been built into the program and included in the construction documents.
In conclusion, archaeology and construction can be friends provided, as in all friendships, mutual understanding and respect prevail.

It has been estimated that almost two thirds of the world's ancient monuments are in Egypt. Discovery of more continues, and will for a long time. It should be kept in mind that ancient Egypt (Pharaonic period) spans about 3000 years. This results in an enormous amount of archaeological remains.

"Archaeology and Civil Engineering Can Be Friends"

R.J. Kachinsky, P.E. F.-ASCE

In early 1990, the massive Greater Cairo Wastewater project was in full swing. A considerable amount of the construction work was in areas containing significant buried archaeological material along the routes of the new sewerage, and was close to the Gizeh Plateau.

All too often large construction projects have been operated under a mandate that says, "If you find anything that looks like it is of archaeological interest, put it in your pocket and keep your mouth shut." This is primarily out of fear that disclosure to the appropriate authorities will result in delays, shutdown and most likely claims. The consequences of this attitude are tragic and beyond measurement

The author, with one foot in the construction/engineering camp and the other in Egyptology wanted to mitigate, if not eliminate, such impacts on the Cairo project.

This involved opening a dialogue with the antiquities people and engaging a full time salvage archaeologist on the construction project. Continuous communication with the antiquities authority was maintained and all relevant information properly recorded all without any delays to the conduct of the construction.

"Civil Engineering at the Gizeh Plateau"

Dr. Mark Lehner

For more than two decades the author has been intimately involved in archaeology at the Gizeh Plateau.

Among these activities has been a base mapping of the plateau which involved establishment of a first order grid and a vertical control bench mark net. This control is still in place and soon will be used in program to undertake a GIS mapping system and relevant data bases. Another activity has been intensive investigations into the condition and means to protect the Sphinx from further deterioration.

Although the Gizeh Plateau is in the desert and about 10 km from the Nile, groundwater levels in the area are fairly close to the surface. Considerable studies have been undertaken to evaluate the impacts of rising groundwater on the Sphinx.

Until fairly recently, most people believed that the limestone blocks which the pyramids of Gizeh are constructed from were quarried on the east bank of the Nile, east of modern Cairo. Extensive study by the author has resulted in a new concept that the quarrying was done at the pyramids site itself. This view is now accepted by most Egyptologists.

At the present time the author is excavating a large area which is exposing large complexes which involve bread making and fish processing as well as other habitations. These facilities are contemporary with the time of the pyramids construction.

At this site are several anomalies in sand and silt stratification which shed light on how the Nile flooding behaved in ancient times to fairly recent time.

In conclusion, sophisticated survey and mapping, ground water evaluation and control and geotechnology all have played and, continue to play a role in my work at the Gizeh Plateau.

"Conservation of the Pyramids and Sphinx: Need for Civil Engineering"

Dr. Zahi Hawass

The magnificent monuments at Gizeh have been assaulted by forces of nature with assistance by man almost since they were constructed, nearly 5000 years ago. In the last century man's assistance has accelerated greatly. A prime example being air pollution from automobiles.

Very recently, a proposal to construct a major highway just south of the pyramids was at first approved and later was cancelled at the last minute.

Ongoing subsurface exploration is hampered by high groundwater in the area. This necessitates complex dewatering efforts and most importantly, disposal of the pumped water without creating problems at adjacent monuments.

In October 1992 the Cairo area was hit by a significant earthquake. Fortunately no major damage resulted at the Gizeh monuments, but Egypt is subject to fairly frequent seismic events. There is a pressing need to make seismic evaluations of the area and develop mitigation programs.

Moving and accommodating a large and growing amount of tourists each year requires "monument friendly" modes of transport on the site as well as provision of sanitary facilities and safe disposal of the waste material.

Erosion of the monuments from sand and the constant need to cope with the sand blowing out of the adjacent desert calls for interventions which will mitigate but not compromise the vistas.

In conclusion, the monuments are both massive and fragile. They require continuous attention to prevent them from suffering irreversible damage. Civil Engineering inputs are vital to these efforts.

Phillips Winthrop House: Home of The Engineering Center and BSCES/ASCE

by
H. Hobart Holly, P.E.[1]

Foreword by
Anni H. Autio, P.E., M.ASCE[2]

Foreword

Phillips Winthrop House: Home of TEC and BSCES/ASCE, One Walnut Street, Boston MA

The civil engineering profession in Boston has a very rich and proud heritage. In 1990, the Boston Society of Civil Engineers Section of the American Society of Civil Engineers (BSCES/ASCE) became one of three sponsoring engineering societies to establish The Engineering Center (TEC) at the Phillips Winthrop House located in the historic Beacon Hill area of Boston, just two blocks from the Massachusetts State House. The other sponsoring engineering societies that constitute TEC are the Massachusetts Association of Land Surveyors and Civil Engineers (MALSCE) and the American Consulting Engineers Council (ACEC) of Massachusetts (formerly ACEC/New England).

TEC is a consortium of associations and professional societies that provides educational programs and information services to engineers, land

[1]H. Hobart "Hank" Holly served as BSCES/ASCE's History and Heritage Committee Chair for many years. Hank passed away in November 1996.

[2]Project Manager, Camp Dresser & McKee Inc., 10 Cambridge, Center, Cambridge, MA 02142 (1997-1998 BSCES/ASCE President and 150th Anniversary Chair for BSCES/ASCE)

surveyors, related professionals and the public. The heritage at TEC is a reflection of the people and organizations who comprise it. The building itself serves as a living monument to our profession embodying the best of the past, present, and future contributions that our engineering community makes to society. TEC's mission is guided by the visions of the sponsoring organizations and its founding members.

Commemorating BSCES's 150th anniversary year celebration, an exclusive history and heritage dinner is planned at TEC for 1998 ASCE Convention attendees. Tales from yesteryear are planned by costumed speakers who will bring forward anecdotes from the past including contributions made by our early society members.

Introduction

The building at One Walnut Street was the product of Charles Bulfinch, the leading architect of his time. While Bulfinch was engaged as the architect for this building as well as for the Massachusetts State House and many other historic buildings, Loammi Baldwin was serving as engineer for the historic Middlesex Canal, and Simeon Borden was establishing the Borden Base Line. Thus, One Walnut Street stands as a tribute to these pioneers in their respective related fields and to all of those who have made contributions to the Commonwealth of Massachusetts and the nation through the professions they represent.

It is therefore appropriate that the Boston Society of Civil Engineers Section/ASCE, the oldest technical society in the United States (established in 1848), have its headquarters in a building that reflects heritage and prestige.

Henceforth, a visit to the Engineering Center will provide an experience that links the endeavors of the present with a past in which the engineers of today can take pride and inspiration.

Early History

One of Charles Bulfinch's most noted achievements as the design of many outstanding homes on Beacon Hill that established a style of elegance that has made the area on the hill famous. The new home of the Engineering Center was built on the corner of Beacon Street and what is now Walnut Street, with its original entrance on Beacon Street. According to the original development, Walnut Street was laid out as a way in 1799 and did not become an accepted street until some years later. Built in 1804 in the "square style" for which Bulfinch was noted, it was one of the earliest, and most probably the first, brick house on Beacon Street on which it was later numbered 38.

The builder and first resident of the house, John Phillips (1770-1823), was one of Boston's most prominent citizens. From 1804, the year the house was built, until his death he was a member of the Massachusetts State Senate, serving as presiding officer for the last ten years. When Boston became a city, Phillips was elected the first mayor in 1822. He served for one year and declined re-election for reasons of health. Born in the house was his son Wendell Phillips, the famous orator for anti-slavery and other causes.

In the 1809 watercolor by J.R. Smith, the Phillips House is shown at the left with the entrance on Beacon Street. Courtesy of the Bostonian Society/Old State House.

Of particular interest, the land deed from the developer, Jonathan Mason, to John Phillips contains the restriction that no building in the development shall be over three stories in height, exclusive of cellar and roof, for a period of thirty years. During the Phillips ownership of nineteen years, there is no record of significant changes having been made to the house.

Two years after John Phillips died in 1823, his heirs sold the house to Thomas Lindall Winthrop, who was also a very prominent Bostonian. He served in the State Senate and as Lieutenant Governor of Massachusetts from 1826 to 1832 while he resided at the house. He was highly esteemed especially for his work on behalf of public schools. He served as president of both the Massachusetts Historical Society and the American Antiquarian Society.

To accommodate his large family, Winthrop moved the entrance to the Walnut Street side and altered the Beacon Street facade. It appears that he raised the upper story and made the third-story windows higher, thus enlarging the living space while remaining within the three story restriction. The ells were added in Winthrop's time and changes were made to the staircase and the interior. Since Bulfinch was still living at this time, it is possible that he might have had a hand in the alterations.

On Winthrop's death in 1841, the house was sold at public auction to Thomas Dixon. Thomas Dixon brought international fame with him. He was born in London but served the Dutch government as Consul General at Boston and abroad. He was made a Knight of the Netherlands for his services.

This 1843 view shows the Phillips-Winthrop House with the entrance on Walnut Street. Courtesy of the Bostonian Society/Old State House.

During the time of Dixon's ownership, the Boston Society of Civil Engineers was founded not too far away at the United States Hotel in Boston on April 26, 1848. A few months later, on July 3rd, the first regular meeting of the society was held and a room for meetings was found in Joy's Building on Washington Street just a short walk away from the house.

Dixon's family owned the house until 1858 when they sold it to Nathan Matthews. Since Matthews all owned buildings at Nos. 3 and 5 Walnut Street, he made some changes to the house on the side that faced his other properties. Matthews was a self-made businessman who was noted for his philanthropies. He donated Matthews Hall and several scholarships to Harvard.

Matthews lived in the house only two years and then sold it to John Chipman Gray who owned it for but one year. Gray was a prominent lawyer. He served as Judge Advocate, and for five years was a lecturer at Harvard Law School.

Robert M. Mason purchased the house in 1861 but appears not to have taken up residence there until 1866. Mason was a successful Boston businessman whose principal philanthropic interest was the Massachusetts Soldiers Fund. Mason's daughters, Ida M. and Ellen E Mason, inherited the house in 1879 and lived there together for about fifty years. The sisters were prominent socially in Boston and at summer places in Newport, Rhode Island, and Dublin, New Hampshire. It was Mason who added the building's mansard roof, effectively making it a four-story building, the three-story habitation having long expired. Also added were some Italianate features that may have reflected the Mason family's residence abroad for some years.

Recent History

In 1931 Mrs. James J. Storrow purchased One Walnut Street from the Mason Estate, and in 1939 donated it to the Judge Baker Foundation as its headquarters. The work performed to transform the former private residence to accommodate institutional functions removed some of Bulfinch's and other later architectural features from the interior, but made it adaptable to the uses of the Engineering Center without requiring extensive alterations. Under this ownership a number of charitable and service organizations had activities in the building, but by far the most important were the Judge Baker Guidance Center and the Boston Children's Services Association.

In 1976 the building was acquired by the Phillips-Winthrop House Trust to serve as the law offices of Mahoney, Hawkes & Goldings. Some interior changes were effected in order to suit its new, more modern function. In 1978 an exterior restoration was accomplished that removed many decorative features that had been added over the years. Most noticeable was the removal of its exterior gray paint in order to expose once more the old red brick. On June 7,1990, the Phillips-Winthrop House Trust sold One Walnut Street to The Engineering Center Education Trust (TECET).

Heritage

The building at One Walnut Street that one views today is an historic building that contributes to, and is part of, historic Beacon Hill. The mansard roof is certainly not Bulfinch, but even with this later addition the Bulfinch lines and features are still in evidence. The changed features that remain reflect the history of Beacon Hill and Boston over the nearly two centuries since the house was built. The interior arrangement is much altered except in the staircase and fireplace,. The proportion, of the rooms and much of the original woodwork has been retained, preserving an atmosphere of substance and elegance culled from the past. In this building, engineers can be reminded of their rich heritage on which thev base their present and future.

As The Engineering Center, history is being made at One Walnut Street by the engineers associated with it. Although actual engineers will not become part of the history of the house, their actions will merge into the lives and dreams of the past residents of the house who are the heritage that we inherited.

One Walnut Street represents a proud past and is a fitting remainder to those of the engineering societies that constitute The Engineering Center of the great heritage that they represent as they make history today and in the future.

NOTE - This article was first published by BSCES/ASCE under the title "The Engineering Center: 1 Walnut Street, Boston" in Fall 1990 issue of Civil Engineering Practice: Journal of the Boston of Civil Engineers Section/ASCE, Volume 5, Number 2.

REFLECTIONS ON ASCE'S HISTORICAL LANDMARK PROGRAMS

Alan L. Prasuhn, FASCE, PE[1]

Abstract

The International Historic Civil Engineering Landmarks are one component in a history and heritage program of the American Society of Civil Engineers that dates back to 1964. During the past 21 years, ASCE has joined with engineering societies in 22 other countries to recognize significant international engineering landmarks. At present, ASCE have recognized 28 such projects. The paper provides a perspective on many of these landmarks and their relationship to some of the 170 National Historic Civil Engineering Landmarks.

As with most of the ASCE history and heritage programs, there are two primary purposes for this activity. The first is to instill in civil engineers, an increased pride in, and a greater appreciation of, their chosen profession. The second purpose is to make the general public aware of the importance of civil engineering and the role civil engineers play in improving their daily lives. In addition, the international recognition process provides a unique opportunity to bring the world-wide civil engineering profession closer together.

Introduction

The ASCE History and Heritage program was authorized by the ASCE Board of Direction in 1964. Shortly thereafter, Neal FitzSimons became the chairman of the Committee on the History and Heritage of American Civil Engineering (CHHACE) and continued to serve in this position for almost 25 years. The author took over from him in 1990 and chaired the committee until a year ago at which time Jerry Rogers took over.

The primary purposes for the program are to (1) create an awareness and pride in civil engineers for the history and heritage of their chosen profession; and (2) inform the

[1] Chairman, Department of Civil Engineering, Lawrence Technological University, 21000 West Ten Mile Road, Southfield, MI 48075, (248) 204-2549.

general public of the role that civil engineering has played in the development of the country and in improving the quality of life.

The landmark programs have been a vital part of the history program, although there are many other components including oral history, history publications, audio-video presentations, and convention sessions. The National Historic Civil Engineering Landmark (NHCEL) program dates back to the early days of the ASCE history program and the recognition of the Bollman Truss Bridge at Savage, Maryland in 1966 as ASCE's first National Historic Civil Engineering Landmark. This program has continued to grow and flourish until ASCE now has approximately 170 NHCEL's from coast to coast. The similar International Historic Civil Engineering Landmark (IHCEL) program is much younger dating from 1979. There are now 28 IHCEL's in some 22 countries around the world. [CHHACE 1992]

The American Society of Civil Engineers will shortly celebrate 150 years of American Civil Engineering. At that time, a perspective on how civil engineering has evolved over the previous 150 years, and where it is heading will certainly be appropriate. With that thought in mind, this paper attempts to provide a preliminary focus to the landmark programs. In particular, what connection or connections exist among the landmarks and between the national and international landmarks?

The IHCEL's have been nominated for a variety of reasons: CHHACE thought that a particular project would be an appropriate landmark, the local civil engineering society wanted a project so recognized, a project was selected to mark a pending ASCE international tour or presidential visit; to mention just a few. For whatever reason, this rather ad hoc selection process has resulted in a list of landmarks that not only showcase the breadth and depth of civil engineering, but also provide an acknowledged record of our global heritage.

Civil Engineering

Although civil engineering has been around for millennia as some of the landmarks identified in this paper will demonstrate, the term "civil engineering" is far more recent. The first identified use of the term is frequently credited to John Smeaton in the early 1760's [Buchanan 1989]. In 1766 he referred to himself as John Smeaton, Civil Engineer, on the title page of a report to the Forth and Clyde Navigation [Watson 1989]. In addition, he arguably laid the foundation for civil engineering as a profession in the 1760's. It is eminently appropriate that his most famous project, the Eddystone Lighthouse, has been identified as an IHCEL. The citation approved by the ASCE board of Direction reads:

The Eddystone Lighthouse was erected 1756-1759 on the Eddystone Rocks by John Smeaton, the first individual to call himself a civil engineer. The first masonry-tower lighthouse to be built at sea, its form was adopted universally. Removed in 1882 be-

cause of erosion of its foundation, it was partially reconstructed on the Plymouth Hoe in 1884. It remains today a supreme symbol of heroic civil engineering achievement. [All citations are taken from ASCE Reports to the Board of Direction or from CHHACE 1992. They are shown in italics.]

The Eddystone Lighthouse was actually the third lighthouse (but first successful) to be constructed on the Eddystone Rocks. It became the prototype for all future lighthouses subjected to wave forces. Significant lighthouses to follow are the Bell Rock Lighthouse off the east coast of Scotland, and the Minot's Ledge Lighthouse near Cohasset, Massachusetts. The latter, constructed during the period 1855-60, was recognized as a NHCEL because of its design and construction to withstand open-sea wave forces.

The Ecole Nationale des Ponts et Chaussees in Paris, France was founded by Trudaine in 1747. Recognized as an IHCEL by ASCE as the oldest civil engineering school in the world, its graduates have had a major impact on the art and science of civil engineering throughout the world.

Bridges

Not surprisingly, over a third of the IHCEL's are bridges. In all of their manifestations, bridges are undoubtedly the works most identified with civil engineering. While it is frivolous to speculate on the first bridges, early bridge types remain in the form of clapper bridges and even stepping-stones. Roman bridges remain, although none have been nominated as international landmarks. The earliest bridge selected as an IHCEL is the Anji or Zhaozhou Bridge constructed c. 605 AD in the Hopei Province, China. Although very advanced for it time, e.g. European bridges did not have hollow spandrels for over 1000 years; because of its obscurity it did not influence bridge design in the western world. Over the centuries, major European bridges were primarily arch bridges in the Roman tradition. The first major departure, and also ASCE's first IHCEL, was the Iron Bridge in Coalbrookdale, England. The ASCE citation reads:

This bridge, completed in 1779, is recognized as the first iron bridge in the world. Standing today, it is an outstanding international monument to both the civil engineering profession and the industrial revolution.

As the first significant use of metal as a building material, its assembly followed traditional wood and masonry techniques; however, it rapidly led to the development of an entirely new technology. One hundred years of development and experimentation culminated in another IHCEL, the magnificent Forth Railway Bridge located at South Queensferry, Scotland just outside of Edinburgh. Its citation follows:

Built between 1882 and 1890, this British railway bridge held for 27 years the world's record for span (521 meters). To achieve this the internationally famous design team of British civil engineers, John Fowler and Benjamin Baker, Hon. M. ASCE, developed the unique double cantilever profile and utilized mild steel. This magnificent structure remains in full service today as a key link in the British Rail system. The 101-meter tall towers create a shipping clearance of 46 meters. The overall length of the bridge is 2529 meters.

The iron and steel bridge designs of Great Britain were exported to the New World as represented by the 1917 cantilevered Quebec Bridge (designated as an IHCEL by CSCE and ASCE in 1987), and a remote, but spectacular structure on the White Pass and Yukon Railway. The IHCEL citation for the railway, one of three railways designated as IHCEL's, is as follows:

The White Pass and Yukon Railway extending from Skagway, Alaska to White Horse, Yukon Territory, was constructed in only 27 months in 1898-1900 by American and Canadian engineers. The Railroad passed through and across glacial terrain, far removed from supplies in the United States, and represented the first cold region engineered construction in Alaska.

The Sydney Harbour Bridge, in Australia, is another product of the one-time British Empire. Like many other bridges around the world, this imposing structure has become a symbol of the country itself.

Not only the design, but the Victoria Falls Bridge itself was exported during colonial days to span the Zambezi River between what are now the countries of Zimbabwe and Zambia. Constructed just below Victoria Falls, one of the natural wonders of the world, its citation hardly does it justice:

The Victoria Falls Bridge, completed in 1905, is a 152-meter span, steel-lattice, two-hinged arch bridge with a deck level 122 meters above the Zambezi River. Conceived by Cecil Rhodes as a key link in his proposed Cape-to-Cairo railway, it is situated just downstream of the Victoria Falls in a site of unsurpassed grandeur. Although a product of the colonial period, it continues to serve and improve the lives of all peoples living in the region.

Similarly, the skills developed by Gustave Eiffel who has three IHCEL's to his credit (the Ponte Maria Pia in Oporto, Portugal - 1877, the Statue of Liberty - 1886, and the Eiffel Tower - 1889), influenced the design of another IHCEL, the Viaducto del Malleco constructed in Chile in 1890.

No mention has been made of wood or suspension bridges, and, with the exception of the "Bridges of Niagara," none at present, are included in the list of international landmarks. These two bridge forms owe much of their development to American civil engineering, and are consequently well represented in our NHCEL's. Wood covered bridges include the Blenheim Bridge in New York (1855), the Bridgeport Bridge in California (1862), and the Cornish–Windsor Bridge between New Hampshire and Vermont (1866). Early suspension bridges include John A. Roebling's aqueduct (1848) that carried the D&H canal over the Delaware river, but subsequently was converted to vehicular traffic, and Charles Ellet Jr's Wheeling Suspension Bridge (1849). Long-span suspension bridges are represented by Roebling's Brooklyn Bridge (1883), the George Washington bridge (1931), and the Golden Gate (1937).

Much of the development of metal bridges occurred within the United States. This record is documented by a long list of outstanding NHCEL's. Only the most recent landmark designation ceremony will be mentioned, namely the recognition of the 188-m steel arch Navajo Bridge across the Colorado River in Arizona in May, 1998. Located in a spectacular location within the Navajo Nation, its primary significance is the vital role it played as the only crossing of the Colorado River in a length of 600 miles.

It would be remiss not to include mention of the reinforced concrete bridge on the IHCEL list. This is the 1930 Salginatobel Bridge located in the Graubunden Canton in Switzerland. The citation speaks for itself:

Designed by Robert Maillart, the Salginatobel Bridge represents a major innovation of structural type - the three-hinged, hollow-bow arch of reinforced concrete - using a new method of staged-arch construction. Visually unique, this unprecedented form by the most celebrated bridge designer of his time, is also considered a work of art.

A student of Maillart's, Christopher Menn, continues to keep Switzerland in the forefront of concrete bridge design.

Water Resources, Water Supply, and Flood Protection

The above heading provides a second group of IHCEL's. The oldest water-related landmark is the Ifugao Rice Terraces in the Philippines.

Dating from 100 BC, this national cultural treasure is the oldest and most extensive use of terraces in the world. The 20,000 hectares of terraces represent a rearrangement of the Cordillera Mountain Range from bedrock to topsoil. It was built and maintained by communal hard work without slave labor or government dictatorship. The engineering principles of hydrology, sustainable development, and efficient use of water resources and irrigation all are embodied in the careful design of this ancestral land management program.

The Netherlands has fought the encroachment of the sea and attempted to reclaim land for centuries. The most significant historic project is the Zuiderzee Enclosure Dam recognized as ASCE's second international landmark. This structure, built in 1927-1932, has successfully barred the sea for over 65 years. The Zuiderzee closure protects a large area north of Amsterdam, and has allowed construction of polders to reclaim much of the enclosed area from the sea. Recent flood protection works, on an even larger scale, but with less environmental impact, resulted in their recognition as one of the seven modern civil engineering wonders of the world.

Other water supply projects approved as IHCEL's include the Acquedotto Traiano-Paolo in Rome, Italy. The original aqueduct built by the Emperor Trajan, c. 110 AD, along with the other aqueducts and numerous civil works provided the advanced infrastructure of ancient Rome. Largely rebuilt in the 17th Century, it continues to provide water for the fountains of Rome as well as part of the needs of the modern city. It remains a true monument to Roman Civil Engineering. Ruins on a grand scale of other Roman aqueducts likely to qualify as IHCEL's include the Pont du Gard near Nemes, France and the Segovia Aqueduct in Spain. Another international landmark is the Queretaro Aqueduct. This 1280-meter long, 74 semi-circular stone arch aqueduct, completed in 1738, provided a dependable supply of clean water to the city of Queretaro, Mexico. Based on the Roman aqueducts, it remains virtually intact today, and is one of Mexico's most important monuments. With a maximum height of 23 meters, it is a remarkable example of 18th century civil engineering practice in Mexico.

The Snowy Mountain Hydro-Electric Scheme in New South Wales, Australia was recently recognized as an IHCEL. This represents one of many examples where a complex system is utilized to provide multipurpose water supply, flood protection, and hydro-electric power to meet the demands of modern society.

Numerous NHCEL's throughout the United States pay homage to the American civil engineer's expertise in this area. Let it suffice to identify a single project, the Hohokam Canal System in the Salt River Valley near Phoenix, Arizona. This extensive irrigation system, constructed and utilized by the Hohokam Indians between 600 and 1450 AD, foreshadowed by several centuries the important role the modern civil engineer was to play in the development of the West. It is an outstanding example of the modification of the environment for beneficial use.

Transportation

In addition to the White Pass and Yukon Railway, two other rail systems have been recognized as IHCEL's. First is the Dublin-Belfast Rail Link constructed between 1842 and 1855 to provide a vital connection between the two capitals of Northern Ireland and the Republic of Ireland. It is recognized primarily for the 536-m Boyne Bridge and Viaduct, which represented the first large-scale use of wrought-iron latticed girders as well as the first full-scale test of continuous beams. The second is the North Island Main Trunk Railway in New Zealand. Completed in 1908, the railway linked Wellington and Aukland, New Zealand, permitting overland travel and development of the hinterland. Built under challenging conditions and over difficult terrain; cuts, fills, and tunneling were minimized by careful use of the topography and by features such as the famed Raurimu Spiral.

The Panama Canal was designated as an IHCEL in 1984. The citation for this world-class project is as follows:

Originally undertaken by the French, the canal was redesigned and constructed by American engineers between 1903-1914. Combining the skills of sanitary, hydraulic, geotechnical, structural and railroad engineers, and effectively mobilizing the efforts of thousands of workers and the power of diverse machines, the greatest sea-to-sea lock canal of all time was successfully built and remains today a major artery in world trade. The chief engineers of this American project were John F. Wallace, 1900 ASCE president, John F. Stevens, 1927 ASCE president, and George Goethals.

The most recently dedicated IHCEL is the Gota Canal in Sweden. Originally conceived in the mid 1500's, actual construction took place from 1810 to 1832. Thomas Telford, the first president of the Institution of Civil Engineers (England), was consultant to the project.

Two roadway projects are designated as IHCEL's. The first is the extremely remote Lake Moeris Quarry Road located in Egypt's northern Faiyum Desert. Dating from the Old Kingdom period in Egypt (2575-2134 BC), it is recognized as the oldest surviving paved road in the world. It was paved with large slabs of limestone and sandstone, of which approximately 47% of it total length of 11.7 km remains. The second is the Alaska Highway extending from Dawson Creek, British Columbia to Delta Junction, Alaska. This 2500-km highway, built in just eight months in 1942, provided an essential transportation link for troops and equipment to Alaska and Northwest Canada during World War II. The road also involved many pioneering permafrost construction techniques. Since 1942, the highway has remained a major transportation link for both tourist and commercial traffic.

Civil Engineers

The recognition of an historic civil engineering landmark usually venerates the civil engineer as well. The paper commenced with John Smeaton. Other engineers associated with the IHCEL's and identified along the way include John Fowler, Benjamin Baker, Gustave Eiffel, Robert Maillart, and Thomas Telford. The final IHCEL included herewith recognizes the creative engineering genius of Marc Brunel and his son, Isambard Kingdom Brunel. Marc Brunel, a French naval officer, came to the US during the French Revolution, and served for a short period as the City Engineer of New York City. He ultimately settled in England where he successfully developed machinery to mass-produce "blocks" for the sailing ships of the day. His greatest achievement was the Thames Tunnel, and the tunneling shield necessary for soft-ground tunneling. The ASCE citation follows:

Built in the period between 1828 and 1843, this tunnel was the crowning achievement of Marc Brunel and the inception of Isambard Kingdom Brunel's illustrious career. The tunnel opened a new era in tunneling practice. It was the first shield-driven tunnel, the first successful soft ground subaqueous tunnel and, in 1869 was adapted as the first subaqueous railway tunnel.

I.K. Brunel went on to establish the Great Western Railway in England (one of the pioneer railway lines); three steamships, the Great Western, the Great Britain (recently restored in Bristol, England), and the leviathan Great Eastern (which was used to lay the first transatlantic cables); and numerous bridges, many of which remain today.

It was another fifty years before the first successful subaqueous tunnel was opened in North America. This was the St. Clair Tunnel (1888-91) between Port Huron, Michigan and Sarnia, Ontario. This was followed by the more ambitious Hudson and Manhattan Railway Tunnel that was started in 1874 but not completed until 1908. Both are recognized as IHCEL's.

Length limitations have resulted in an emphasis on the international landmarks. Left out as a result are the outstanding American civil engineers who laid the foundations of our profession today. The ASCE Board of Direction has approved a Civil Engineering Hall of Fame. As the Hall of Fame becomes established, it is most appropriate that our attention turn to an increased awareness and appreciation of these gallant civil engineers and the debt we owe to them.

Conclusions

By examination of many of the 28 ASCE International Historic Civil Engineering Landmarks, we see a worldwide pattern of dedication to the improvement of life on this planet. Although not emphasized in the paper, the 22 countries on six continents currently containing landmarks provide a fair global representation of civil engineering achievement. Last, but certainly not least, the recognition ceremonies held in conjunction with the local engineering societies have created a professional good will that one could only wish would be duplicated at the government level.

References

Buchanan, R.A. (1989). *The Engineers – A History of the Engineering Profession in Britain 1750-1914*. Jessica Kingsley Publishers, London, England.

Committee on History and Heritage of American Civil Engineering (1992). *Guide to History and Heritage Programs*. ASCE, New York, NY (but updated).

Watson, Garth (1989). *The Smeatonians – The Society of Civil Engineers*. Thomas Telford Ltd., London, England.

29 YEARS DOCUMENTING ENGINEERING HERITAGE

Eric DeLony[1]

Abstract: *Engineering and industrial heritage is the foundation of subsequent preservation efforts that have transformed communities and the way that people regard engineering works and the industrial workplace. In the United States, HAER has been one of the primary catalysts that has effected this change in American life over the last twenty-nine years. During this era of deindustrialization and infrastructure rehabilitation, places like Paterson, New Jersey, Lowell, Massachusetts, Butte, Montana, Birmingham, Alabama, and the Mon Valley in Pittsburgh - the industrial heartlands of the United States - are striving to redefine their images and urban cores. These new images are based in part on the recognition, appreciation, and when possible, the preservation, continued and adaptive reuse of industrial and engineering fabric.*

Introduction

The Historic American Engineering Record (HAER) was established in 1969 by the Congress so that documentation on outstanding works of engineering and industry could be preserved in the Library of Congress. Even then, many realized that few physical works of engineering or industry could realistically be saved as historic monuments. However, preservation through documentation was possible. The civil engineers of the United States, borrowing a precedent established by architects in 1933, established HAER, a sister program to the Historic American Buildings Survey (HABS), by signing a tripartite agreement along with the Library of Congress and the National Park Service. Signatures included those of Waldo Bowman, President, American Society of Civil Engineers (ASCE), whose national membership serves as advisors; L. Quincy Mumford, Librarian of Congress, where the records are curated and made available to the American public; and George Hartzog, Director, National Park Service (NPS). The NPS maintains a professional staff of architects, engineers, historians, and photographers to prepare the drawings, photographs, and histories that form the national record.[1]

---

1. Chief & Principal Architect, Historic American Engineering Record, National Park Service, 1849 C Street, NW, NC300, Washington, DC 20240, (202)343-4237, eric_delony@nps.gov.

Celebrating a quarter century documenting engineering heritage in 1994, afforded me the opportunity of a retrospective look at the work we had accomplished over the last twenty-five years. I believe we all appreciate how much of the world has been impacted by deindustrialization and other social and economic phenomenon as we approach the new millennium, but I doubt if few of us appreciate that this change is as profound as the industrial revolution of the 18th and 19th centuries. Like-it-or-not, we are experiencing a change from an industrial to a service economy. There is much to be learned from sharing knowledge about America's engineering heritage experience. This paper will reflect on the work of the Historic American Engineering Record by reviewing HAER's efforts to identify, document, and when possible, preserve outstanding engineering works and industrial landscapes.

HAER and the National Park Service have been working to expand heritage memory to include the achievements of engineers and laborers on the vast panoply of the American experience. Established twenty-nine years ago, HAER strives to create a national archive of America's industrial, engineering, and technological accomplishments. Sites recorded serve as the foundation of subsequent preservation efforts that have transformed communities and the way people think of the industrial work place. The steel mills, factories and foundries that helped create the fortunes of industrial magnates and worker livelihoods and the canal, road and rail transportation networks that served these industries are now beginning to be thoughtfully regarded and preserved with new insights.

The goals of this paper are as follows: to summarize the American experience of urban revitalization through industrial and engineering heritage; to examine the governmental context in which industrial and engineering heritage works by outlining the federal, state and private partnership through which American preservation works; and thus describe the mission of the Historic American Engineering Record. Through its federal authority, national standards, its archives at the Library of Congress, the summer recording program and student internship opportunities, HAER, since 1969, has instilled a national ethic that is beginning to recognize the oft-forgotten contributions of its engineers, industrialists, and laborers. Techniques used include heritage site identification, evaluation, interpretation and documentation; history of technology; industrial archeology; preservation planning; and community mobilization.

In 1994, HAER celebrated 25 years documenting America's engineering, industrial, transportation, and technological heritage. That quarter century can be broken down into three epochs:

* 1969-1979, a decade of discovery, invention, and proselytizing a new field of heritage preservation.

* 1979-1984, a period of crisis due to political and governmental

reorganization when HAER and the other preservation programs of the NPS were transferred to a new Heritage Conservation & Recreation Service (HCRS) during the Carter administration. When Jimmy Carter lost reelection in 1980, HCRS was abolished, many of these agencies scrambled to get back into the Park Service. HABS/HAER and most of the former federal preservation programs were reintegrated back into the Park Service.

* 1984-1998, over a decade of incremental growth for HAER and engineering heritage at a rate of 2%-5% per year measured by:

Dollars - $750,000 appropriated annually by the Congress is used to leverage nearly an equal amount ($645,000) from other federal agencies, the States, private preservation groups, and private industry for a total operating program of $1.4 million in 1998 to document engineering and industrial heritage.

Jobs - 2,900 or approximately 100 jobs/year have been created in industrial heritage documentation over the last twenty-nine years. This does not account for the hundreds of jobs held by people working in industrial museums or consulting practices creating mitigatory documentation of industrial and engineering sites, or practicing engineers who work on the rehabilitation of historic bridges and other related structures.

Sites recorded - 6,600.

Records - More than 57,000 photographs, 46,000 data pages, 3,000 sheets of measured and interpretive drawings have been transmitted to the Library of Congress and deposited in the public domain.

Factors of Success

Factors that have contributed to the success of HAER's engineering-heritage documentation program include authority, partnership, national standards, public domain, entrepreneurship, and a national training program. Statutory authority, the all-important foundation of HAER's documentation program, is provided in the *National Historic Sites Act of 1935* that authorized the National Park Service " . . . to secure, collate, and preserve . . . to survey and to make necessary investigations, and for the purpose of the act, to contract and make cooperative agreements, to develop an educational program and service for the purpose of making the records available to the public for which reasonable charges may be made . . . " The *1935 Act* established the documentation mandate by calling for the creation of a national record of America's significant architectural, technological, historical, and cultural achievements.

This basic legislative authority was enhanced in 1966 by the *National Historic Preservation Act* which charged federal agencies with stewardship responsibilities for the historic properties they owned, or might affect in carrying out their day-to-day activities. The law also established a federal/state partnership by creating a National Register of Historic Places, a Historic Preservation Grants Program, and a National Archeological Program that would be jointly conducted by state and federal governments. It created a federal Advisory Council on Historic Preservation to advise the President and the Congress on preservation matters. It funded the National Trust for Historic Preservation to encourage preservation activities in the private sector. Specifically, for engineering and architectural heritage, the *1966 Act* mandated that any building, site, structure, or object listed or eligible for listing on the National Register of Historic Places, were to be adversely impacted by a federally-funded or licensed project, it must be documented to Historic American Engineering Record/Historic American Buildings Survey (HAER/HABS) standards if there were no alternatives to demolition. This mitigatory documentation mandate has produced a continuous flow of records into the HAER collection; however, the sites recorded over the last 10-15 years have tended to be of state and local significance. After more than 30 years of application, few sites of national significance are being destroyed.

Another authority factor, perhaps more subtle in its effect, is that the HAER program has been established at the highest level of government. Of the three branches that balance and shape American statecraft - legislative, judicial, and executive - HAER resides within the executive branch, the Department of the Interior, one of the President's secretariats. This means that the documentation of engineering heritage has significance in the daily operation of federal programs and credibility in the public's eye.

HAER is also a "partnership." In addition to the support it receives from the National Park Service, HAER, through it's "tripartite agreement" has the additional backing of two other prestigious institutions - the Library of Congress and the American Society of Civil Engineers. Recently, under the leadership of the last two chairs of ASCE's Committee on the History and Heritage of Civil Engineering (CHHACE), Allan Prashun and Jerry Rogers, and Curtis Deane, the Director of the ASCE's Research Foundation, HAER has become more closely allied with the Society's daily activities. Through a cooperative agreement between ASCE and the Park Service, HAER has worked with groups within the Society by providing images from its collection for ASCE's annual bridge calendar. Funds have been set aside to establish a "Founders Award" for the best documentation donated to the HAER collection. HAER has consulted with ASCE in establishing the first continuing-education course in engineering heritage preservation - the "Historic Bridge Rehabilitation Workshop" offered for the first time to members at seminars held in Boston and Chicago during 1998. HAER and ASCE jointly produced and published *Landmark*

American Bridges with the Little Brown Publishing Company of Boston and presently are working on a sequel dealing with 19th-century engineering achievements. HAER sits on the CHHACE Committee in an *ex officio* capacity.

HAER is the "National Standard" of engineering and industrial heritage documentation. Both a performance and quality standard, it is the means by which all documentation undertaken in the United States is measured. Documentation prepared for the HAER collection is created to a 500-year archival standard and deposited in the Library of Congress.[2] By being part of the Library's collection, HAER documentation is in the "public domain." The Library is responsible for the curatorial maintenance and facilitating the collections availability to the American public. Materials from the collection can be used without restriction other than a courtesy of a credit line identifying the delineator, photographer, or author, and the Historic American Engineering Record.

While the Smithsonian often is referred to as the "nation's attic," I characterize the drawings, photographs, and histories that comprise the HAER collection as the "grey cells" that make up the "national memory" of engineering and industrial achievements. The Library of Congress, where the collection is reposited is the "national treasure chest." Viewed in this context, the act of documentation is a powerful tool. The work we do is fundamental, often serving as the foundation from which subsequent preservation efforts are built. When viewed in terms of a "permanent collection" designed to last not five, not fifty, but five hundred years, one can appreciate the profundity of the collection and its compilation.[3]

HAER is "entrepreneurial" in the sense that it can receive funding outside its annual appropriation from other federal agencies, state and local governments, private preservation groups, and equally important, individuals and private industry. Not relying exclusively on federal funding gives the program great flexibility in conducting its business. Shared funding is based on the premise that all sectors of society - government, business, industry and individuals - must participate in a national preservation effort if the historic sites that we enjoy - be they buildings, bridges, canals, or cultural landscapes - have any chance of being saved. Participation, especially financial, encourages our partners to buy into a concept of industrial heritage documentation, and by extension, a commitment to fostering amenity and the quality of life in the United States by preserving the more significant physical attributes of its built and engineered environment.

The summer recording program is the "heart" of the HAER program. Student engineers, architects, historians, industrial designers, archeologists - anyone who can contribute to the compilation of drawings, photographs, and histories - are offered internship opportunities to work on HAER projects during the non-academic portion of the year. Every summer, since its inception, more than a thousand young people have had the opportunity of a "hands-on" experience documenting the nation's industrial, engineering, and architectural heritage. HAER's summer recording program is extremely popular with colleges and universities, the Congress, professional groups and the public, and the clients paying for the services. HAER documentation maintains traditional skills such as field reconnaissance and measuring, delineation, historical research and writing, and large-format photography. The "core" of the summer recording program, and the basic day-to-day philosophy of HAER documentation is the multi-disciplinary team approach.

While some may view documentation as "pedantic," a nice thing to do but not absolutely necessary, I can tell you that having completed several hundred projects over the last quarter century for several hundred clients, that documentation, carefully orchestrated, is a powerful tool. HAER has implanted an ethic of industrial heritage within the American public and recognition that the oft-forgotten engineering structure or industrial workplace has significance and meaning in the vast panoply of the American experience. Engineering and industrial heritage has the potential of involving our working-class citizens in the preservation movement. By recognizing the industrial workplace, we send the message that the occupations of steel, textile, and workers of all types have value. Hopefully, through this recognition we can enfranchise this segment of the population to join in the effort to create amenity through an appreciation of historic places, be they architectural, industrial, or vernacular, and thus enhance the quality of life in America.

Site Selection

A question often asked is how does HAER select sites for recording? Three factors help shape site selection. First, there is a series of thematically selected sites or structures that are threatened; second are the interests of the HAER staff; and third, is filling the gaps in the collection. Infrastructure rehabilitation and deindustrialization have driven HAER documentation over the last quarter century. The need to upgrade our highways and public works - water, sewage, and hydroelectric generating plants - has placed great pressure on recording these sites before they are altered or destroyed. Most of these sites were built in the last quarter of the 19th or first quarter of the 20th-century and while not all retain historic characteristics or machinery, some do and therefore qualify for recording. Rebuilding the nation's highway and rail infrastructure has placed historic bridges in jeopardy. Consequently, more than a thousand bridges have been added to the collection. HAER is in its tenth season documenting national park roads such as the Going-to-the-Sun Highway in Glacier, and historic parkways such as the Merritt in Connecticut and the Columbia River Gorge in Oregon. Deindustrialization has removed the steel industry with its characteristic furnaces and mill buildings from the American landscape. The magnificent mills, blast furnaces, and coke plants along the Monongahela River in Pittsburgh, that for over a century was the world's image of America's industrial might, no longer exist. The same holds true for New England's textile industry, the anthracite regions of northeastern Pennsylvania, and the hard-rock mining fields of the American West. Even the nuclear industry is undergoing rationalization. The march of progress has placed great pressure on the documentation mandate resulting in more work than possibly can be handled in a lifetime.

The criterion for selecting sites for documentation does not depend exclusively on threats and the test of national significance, but also on available funding. The mitigatory documentation program mentioned earlier insures that sites of state and local significance impacted by federal programs are recorded. This allows HAER to focus its limited efforts and to seek funding for documenting nationally-significant sites that may fall through the mitigatory-documentation safety net. Gaps in the collection are the last factor that shapes documentation priorities. After a quarter century, the collection is incredibly rich and comprehensive. None-the-less, there remain areas that are not adequately represented such as the chemical and oil industries, several other extractive industries such as lumber, coal, and gold mining, and even contemporary industries where change occurs so rapidly such as computers and microchips.

Recent Developments

A recent phenomenon that holds great promise for the future is the concept of industrial heritage areas. As former thriving communities become depressed when industries relocate, deindustrialization has stirred local Congressmen to help identify alternative employment opportunities for their constituents. One alternative is *heritage tourism.* Being the national experts on engineering and industrial heritage documentation, HAER usually is called in to help identify not only what is significant, but more important, to identify what is not so derelict sites of limited architectural or technological value can be removed to make way for new construction. Not every vestige of the industrial landscape can be preserved, but there is little argument that some memory, some aspect of steel making or textile manufacturing should be commemorated.

Setting aside selected sites for future interpretation as industrial museums is one preservation option. Others include finding alternative and new uses for industrial buildings and engineering works. Both strategies have worked successfully in American cities to varying degrees, but the most promising and comprehensive concept to emerge in recent years is the industrial heritage area or corridor. Presently, there are about a dozen of these areas in various stages of planning and development throughout the United States. The model for the industrial heritage area was Lowell, Massachusetts, an important location of the American textile industry and one of the first "birthplaces" of the American industrial revolution. Lowell began as an industrial heritage area that, in 1976, evolved into one of the 376 units of the National Park Service . Though operated under a federal, state and local partnership, Lowell is administered by the National Park Service. However, not all sites, no matter how worthy, can be owned or administrated by the National Park Service. The solution for these other sites rests with the creation of local and state partnerships that are responsible for administering a heritage area commission with the possibility of technical assistance from the Park Service. Sites falling into this category include the Southwestern Pennsylvania Heritage Preservation Commission, formerly America's Industrial Heritage Park (AIHP), that encompasses nine counties in southwestern Pennsylvania focusing on the steel industry of Pittsburgh and the affiliated industries that supported it - railroads, mining, and coke-making. Another is the Lehigh and Delaware Navigation Canal National Corridor, an area more than 150 miles in length representing industrialization along the Lehigh and Delaware Rivers that flow from the anthracite coal fields in northeastern Pennsylvania to the former "workshop of the world" - Philadelphia. The Blackstone River Valley National Heritage Corridor, a 50-mile stretch in Rhode Island and Massachusetts, seeks to preserve remnants of the southern New England textile industry. The valley is characterized by tightly-knit 19th-century mill hamlets that dot the countryside along this former industrial and transportation corridor. In the southern United States, the citizens of Birmingham, Alabama, have saved the Sloss Furnaces and now are working to save and interpret related aspects of the iron-making, coal, and mining industries of the "Birmingham District" and its surrounding five-county area. In Augusta, Georgia, the focus is the Augusta Canal and remnants of textile mills and related industries at the falls of the Savannah River.[4]

The encouraging thing about industrial heritage areas is that for the first time, a mechanism has been established where amenity in the form of natural, recreational, and cultural resources is weighed and evaluated equitably with the other forces driving development such as retail sales, housing, office space, parking lots and pavement. In many communities, citizens insist that these values be included in any redevelopment scheme. Community leaders cannot deny that these values are not worthy of preserving, and if nonexistent, are creating them when possible. Features and attributes, both manmade and natural, that allow us to distinguish one community from another, that offers identity and a sense of place, are becoming more and more the primary ingredients that planners and developers use to revitalize communities. The American public has become more sophisticated realizing that the destruction of old buildings and the eradication of waterfronts with seamless shopping malls and suburban neighborhoods no longer are the only options.

Conclusions

Those of us involved in historic preservation use documentation for more than just a "permanent record." When people ask what is the value of documentation, I answer that we are in the amenity business, the responsibility of saving the very best handed down to us so we can pass it onto the future. Saving structures of fine materials, humanly-scaled proportions, notable craftsmanship, and varied textures enhances the quality of life and maintains familiar surroundings. In places where historic architectural and cultural resources are lacking, attitudes supporting good community design may also be absent. Such values are especially needed in America, where we tend to throw away the past, build the expedient, pursue the quick profit, and, in the process, trash the countryside. In this age of instant gratification, suburbanization, and disintegration of our urban cores, engineering heritage and the historic built environment provide a link with that past as well as deeper insights and appreciation of the human imagination and activity portrayed.

Engineers only recently have begun to share this vision of the future development of America. Forums such as this, the annual meeting of the American Society of Civil Engineers, offer the opportunity to inform engineers of their past achievements so they can make choices based not only on the "bottom line," as has been the practice for nearly a generation, but also on aesthetics and quality of life issues as well. Historians and preservationists seek to understand the past, not only to describe it, but because it is one of the only ways to understand the present, let alone to dimly envisage the future. Historians, on the other hand, frequently lack the technical expertise to describe complex engineering phenomena adequately. So much of our historic built environment was built by engineers that its careful maintenance and preservation require the engineer's expertise and insights. Not the engineer with the latest computer graphics and pat-solutions to problem solving, but a "new specialist" like the bridge engineer of the late-19th century-one who is familiar with the materials, mathematics, theories, and the mind set of his or her predecessor so they can craft solutions that respect the material character and human qualities we all enjoy and relish. Many challenges confront the modern engineer, but one that needs immediate attention is the education of a "new specialist" who can learn from the past and adapt for the future.

Endnotes

1. In 1987, the original tri-partite agreement was expanded by a *Protocol* to include the other "founding" engineering societies: American Society of Mechanical Engineers (ASME), Institute of Electrical and Electronics Engineers (IEEE), American Institute for Chemical Engineers(AICE), and the American Society of Mining, Metallurgical and Petroleum Engineers.

2. See the "Secretary of the Interior's Standards & Guidelines for Architectural & Engineering Documentation," *Federal Register,* Volume 48, Number 190, Thursday, September 29, 1983, p. 44,730-44,734.

3. An exciting recent development that will revolutionize the use of the HAER collection is the National Digital Library. The Library of Congress, with the bipartisan support of the US Congress, the Executive Branch and America's entrepreneurial and philanthropic leadership, is creating the National Digital Library. Because the HABS and HAER collections are among the largest and most heavily used in the Library, they will be one of the first to go on-line as part of the American Memory Collections.

4. National Heritage Areas as of 1998 include: Illinois & Michigan Canal (1984), Blackstone River Valle (1986), Delaware & Lehigh Canal (1988), Southwestern Pennsylvania (1988), Cane River in Louisiana (1994), Quinnebaug & Shetucket River Valleys in Connecticut (1994), America's Agricultural Heritage Partnership in Iowa (1996), Augusta Canal in Georgia (1996), National Coal Heritage in West Virginia (1996), Essex National Heritage Area in Massachusetts (1996), Hudson River Valley (1996), Ohio & Eri Canal (1996), South Carolina (1996), Steel Industry in Pittsburgh (1996), and the Tennessee Civil War Heritage Area (1996).

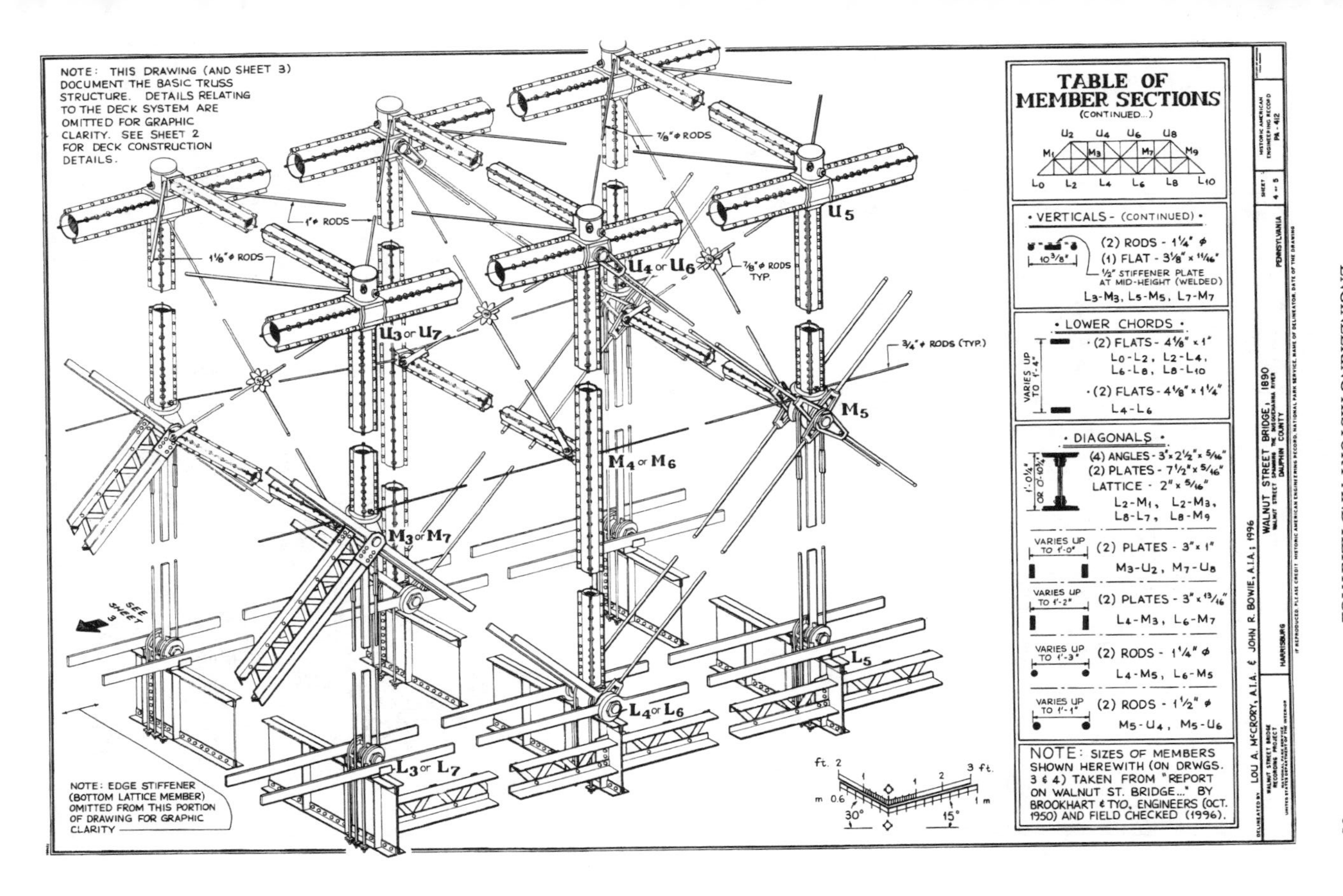
NOTE: THIS DRAWING (AND SHEET 3) DOCUMENT THE BASIC TRUSS STRUCTURE. DETAILS RELATING TO THE DECK SYSTEM ARE OMITTED FOR GRAPHIC CLARITY. SEE SHEET 2 FOR DECK CONSTRUCTION DETAILS.
⅞"φ RODS
1"φ RODS
1⅛"φ RODS
⅞"φ RODS TYP.
¾"φ RODS (TYP.)
U5
U4 or U6
U3 or U7
M5
M4 or M6
M3 or M7
L5
L4 or L6
L3 or L7
SEE SHEET 3
NOTE: EDGE STIFFENER (BOTTOM LATTICE MEMBER) OMITTED FROM THIS PORTION OF DRAWING FOR GRAPHIC CLARITY
ft. 2 1 0 1 2 3 ft.
m 0.6 0 1 m
30° 15°
TABLE OF MEMBER SECTIONS
(CONTINUED...)
U2 U4 U6 U8
M1 M3 M7 M9
L0 L2 L4 L6 L8 L10
• VERTICALS - (CONTINUED) •
10⅜"
(2) RODS - 1¼" φ
(1) FLAT - 3⅛" x 11/16"
½" STIFFENER PLATE AT MID-HEIGHT (WELDED)
L3-M3, L5-M5, L7-M7
• LOWER CHORDS •
VARIES UP TO 1'-4"
(2) FLATS - 4⅛" x 1"
L0-L2, L2-L4, L6-L8, L8-L10
(2) FLATS - 4⅛" x 1¼"
L4-L6
• DIAGONALS •
1'-0¼" OR 0'-10¾"
(4) ANGLES - 3" x 2½" x 5/16"
(2) PLATES - 7½" x 5/16"
LATTICE - 2" x 5/16"
L2-M1, L2-M3, L8-L7, L8-M9
VARIES UP TO 1'-0"
(2) PLATES - 3" x 1"
M3-U2, M7-U8
VARIES UP TO 1'-2"
(2) PLATES - 3" x 13/16"
L4-M3, L6-M7
VARIES UP TO 1'-3"
(2) RODS - 1¼" φ
L4-M5, L6-M5
VARIES UP TO 1'-1"
(2) RODS - 1½" φ
M5-U4, M5-U6
NOTE: SIZES OF MEMBERS SHOWN HEREWITH (ON DRWGS. 3 & 4) TAKEN FROM "REPORT ON WALNUT ST. BRIDGE..." BY BROOKHART & TYO, ENGINEERS (OCT. 1950) AND FIELD CHECKED (1996).
DELINEATED BY: LOU A. McCRORY, A.I.A. & JOHN R. BOWIE, A.I.A.; 1996
WALNUT STREET BRIDGE, 1890
WALNUT STREET SPANNING THE SUSQUEHANNA RIVER
HARRISBURG
DAUPHIN COUNTY
PENNSYLVANIA
SHEET 4 of 5
HISTORIC AMERICAN ENGINEERING RECORD
PA-412

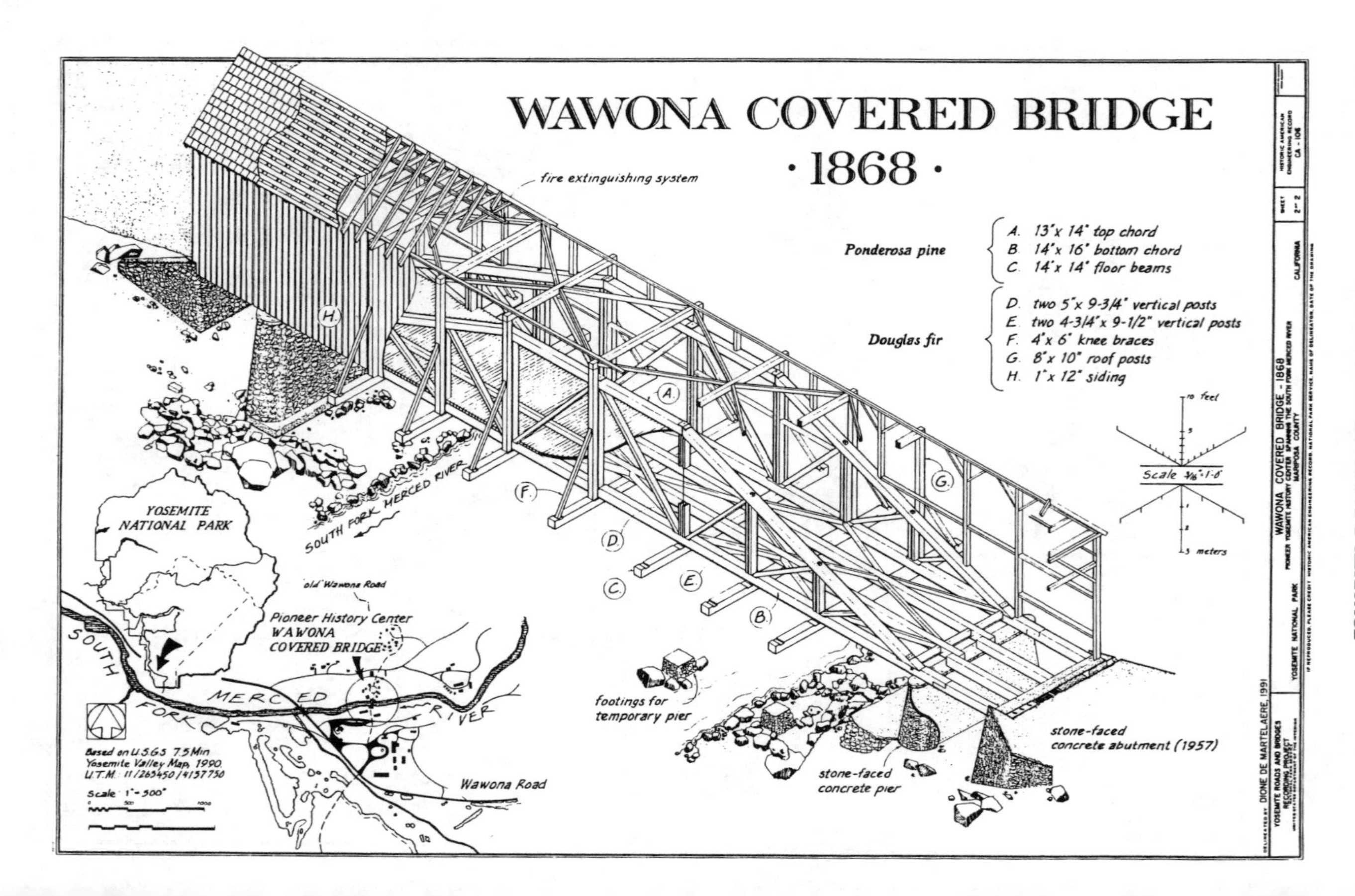

WAWONA COVERED BRIDGE
· 1868 ·
fire extinguishing system
Ponderosa pine
A. 13" x 14" top chord
B. 14" x 16" bottom chord
C. 14" x 14" floor beams
Douglas fir
D. two 5" x 9-3/4" vertical posts
E. two 4-3/4" x 9-1/2" vertical posts
F. 4" x 6" knee braces
G. 8" x 10" roof posts
H. 1" x 12" siding
10 feet
3 meters
stone-faced concrete abutment (1957)
stone-faced concrete pier
footings for temporary pier
SOUTH FORK MERCED RIVER
YOSEMITE NATIONAL PARK
old Wawona Road
Pioneer History Center
WAWONA COVERED BRIDGE
SOUTH FORK MERCED RIVER
Wawona Road
Based on U.S.G.S. 7.5 Min Yosemite Valley Map, 1990
Scale 1" = 500'
DELINEATED BY DIONE DE MARTELAERE, 1991
YOSEMITE ROADS AND BRIDGES
WAWONA COVERED BRIDGE - 1868
YOSEMITE NATIONAL PARK
MARIPOSA COUNTY
CALIFORNIA
CA - 106

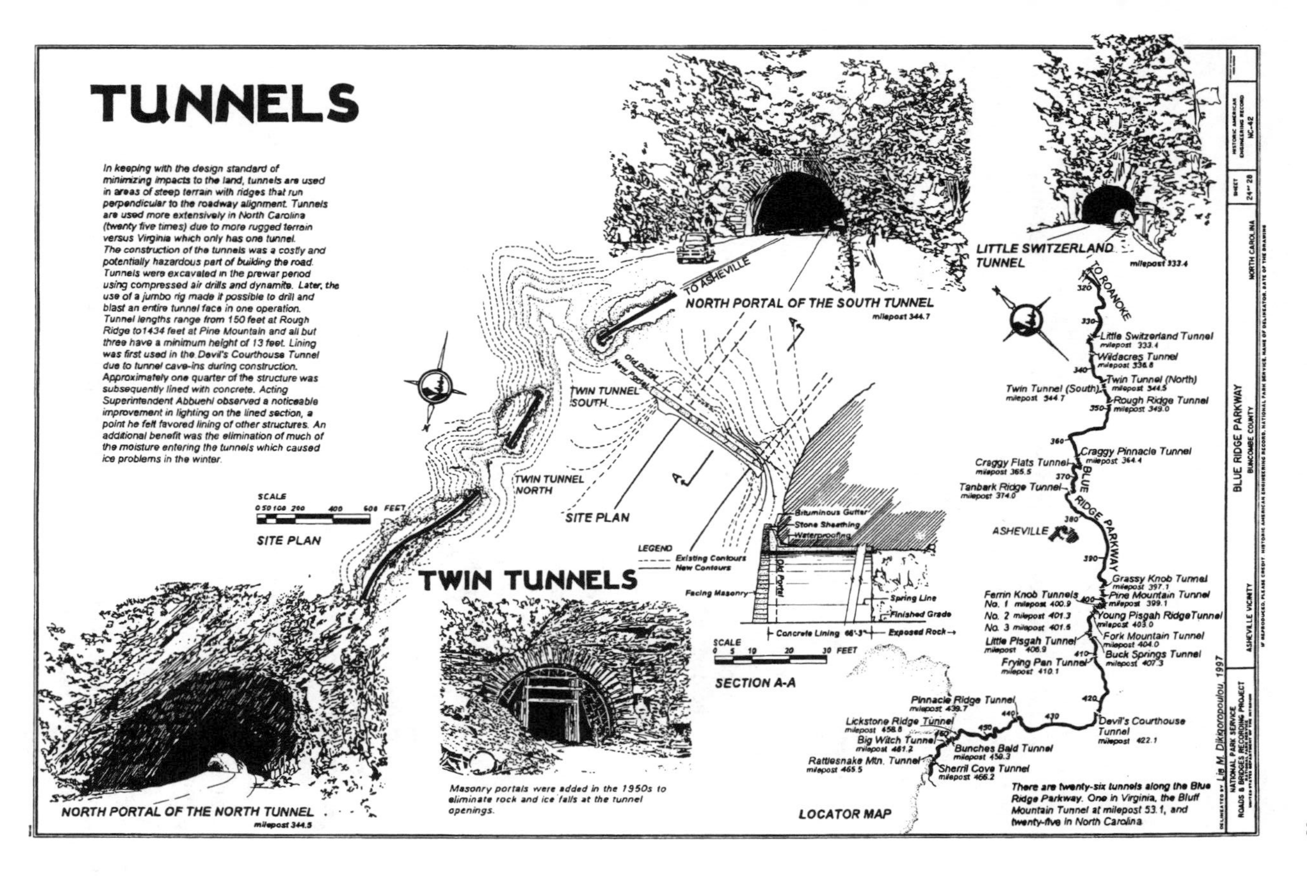
TUNNELS
In keeping with the design standard of minimizing impacts to the land, tunnels are used in areas of steep terrain with ridges that run perpendicular to the roadway alignment. Tunnels are used more extensively in North Carolina (twenty five times) due to more rugged terrain versus Virginia which only has one tunnel.
The construction of the tunnels was a costly and potentially hazardous part of building the road. Tunnels were excavated in the prewar period using compressed air drills and dynamite. Later, the use of a jumbo rig made it possible to drill and blast an entire tunnel face in one operation.
Tunnel lengths range from 150 feet at Rough Ridge to1434 feet at Pine Mountain and all but three have a minimum height of 13 feet. Lining was first used in the Devil's Courthouse Tunnel due to tunnel cave-ins during construction. Approximately one quarter of the structure was subsequently lined with concrete. Acting Superintendent Abbuehl observed a noticeable improvement in lighting on the lined section, a point he felt favored lining of other structures. An additional benefit was the elimination of much of the moisture entering the tunnels which caused ice problems in the winter.
NORTH PORTAL OF THE SOUTH TUNNEL
milepost 344.7
TO ASHEVILLE
LITTLE SWITZERLAND TUNNEL
milepost 333.4
TWIN TUNNEL SOUTH
TWIN TUNNEL NORTH
Old Portal
New Portal
SCALE
0 50 100 200 400 600 FEET
SITE PLAN
SITE PLAN
LEGEND
Existing Contours
New Contours
TWIN TUNNELS
Bituminous Gutter
Stone Sheathing
Waterproofing
Old Portal
Facing Masonry
Spring Line
Finished Grade
Concrete Lining 66'-3"
Exposed Rock
SCALE
0 5 10 20 30 FEET
SECTION A-A
NORTH PORTAL OF THE NORTH TUNNEL
milepost 344.5
Masonry portals were added in the 1950s to eliminate rock and ice falls at the tunnel openings.
TO ROANOKE
Little Switzerland Tunnel milepost 333.4
Wildacres Tunnel milepost 336.8
Twin Tunnel (North) milepost 344.5
Twin Tunnel (South) milepost 344.7
Rough Ridge Tunnel milepost 349.0
Craggy Pinnacle Tunnel milepost 364.4
Craggy Flats Tunnel milepost 365.5
Tanbark Ridge Tunnel milepost 374.0
BLUE RIDGE PARKWAY
ASHEVILLE
Grassy Knob Tunnel milepost 397.1
Pine Mountain Tunnel milepost 399.1
Ferrin Knob Tunnels
No. 1 milepost 400.9
No. 2 milepost 401.3
No. 3 milepost 401.5
Young Pisgah RidgeTunnel milepost 403.0
Fork Mountain Tunnel milepost 404.0
Little Pisgah Tunnel milepost 406.9
Buck Springs Tunnel milepost 407.3
Frying Pan Tunnel milepost 410.1
Devil's Courthouse Tunnel milepost 422.1
Pinnacle Ridge Tunnel milepost 439.7
Lickstone Ridge Tunnel milepost 458.8
Big Witch Tunnel milepost 461.2
Bunches Bald Tunnel milepost 459.3
Rattlesnake Mtn. Tunnel milepost 465.5
Sherril Cove Tunnel milepost 466.2
There are twenty-six tunnels along the Blue Ridge Parkway. One in Virginia, the Bluff Mountain Tunnel at milepost 53.1, and twenty-five in North Carolina.
LOCATOR MAP
320
330
340
350
360
370
380
390
400
410
420
430
440
450
460
DELINEATED BY Lia M. Dikigoropoulou, 1997
NATIONAL PARK SERVICE
ROADS & BRIDGES RECORDING PROJECT
ASHEVILLE VICINITY
BLUE RIDGE PARKWAY
BUNCOMBE COUNTY
NORTH CAROLINA
SHEET 24 of 28
HISTORIC AMERICAN ENGINEERING RECORD
NC-42

ASCE Guide to History and Heritage Programs

Jerry R. Rogers, Ph.D., P.E.[1]

Abstract

The recently revised July 1998 "Guide to ASCE History and Heritage Programs" is available from ASCE, 1015- 15th Street, N.W., Suite 600, Washington, D.C. 20005-2605. The ASCE Guide contains listings of ASCE National and International Historic Civil Engineering Landmarks and Historic Civil Engineering Works, a chronology of American civil engineering, various civil engineering history and heritage programs, and engineering history resources.

Introduction

The "Guide to ASCE History and Heritage Programs" has been recently revised by ASCE Staff Members: Gayle Fields and Jane Howell (Committee Staff Contact) and reviewed by members of the Committee on the History and Heritage of American Civil Engineering (CHHACE) such as Alan Prasuhn- Past Chair, Jerry Rogers- Chair and Edward Robinson. The colorful, informative Guide was dedicated to Herbert Ridge Hands, 1926- 1997, past CHHACE staff contact, who (with Neal FitzSimons' leadership) was instrumental in the establishment of CHHACE, the landmark program, and the first Guide. Eric DeLony (Director, Historic American Engineering Record) and Robie Lange of the National Park Service helped provide NPS listings.

Recently, Neal FitzSimons and Frank Griggs, Jr. have developed a chronology of American Civil Engineering which is being placed on the internet as part of the ASCE educational program directed by Mike Kupferman and funded by the Sloan Foundation. Frank Griggs, Jr. has developed a new video: "On the Shoulders of

[1] Chair- ASCE Committee on the History and Heritage of American Civil Engineering, Department of Civil and Environmental Engineering, University of Houston, Houston, Texas 77204

Giants" which will be distributed soon by ASCE. ASCE thanks all those who developed the early to recent civil engineering history and heritage guides and educational engineering history programs.

Contents of the Guide to ASCE History and Heritage Programs

The Guide to ASCE History and Heritage Programs has alphabetical and state or country listing of National and International Historic Civil Engineering Landmarks and Historic Civil Engineering Works. There are also notations of state landmark projects documented by the Historic American Engineering Record and notations of projects designated by the National Park Service as a National Historic Landmark.

A Chronology of American Civil Engineering begins with the extensive irrigation system constructed by the Hohokam Indians at Salt River, Arizona in 600 and the 1500s with the El Camino Real- The Royal Road, Eastern Branch in Texas to Florida and Western Branch to provide support, defense, commercial ties and political stability for early colonists.

ASCE History PROGRAMS include the ASCE History and Heritage Award, Special ASCE Recognitions, History Committee/ Presidential Award, Historic Civil Engineering Landmark Program, Landmark Designation Procedure, Local/State Landmark Designation Procedure, 13 Helpful Suggestions for State and Local History and Heritage Committees, and the ASCE Oral History Program.

Among the RESOURCES listed are the Cooperative Agreements; Historic American Engineering Record (HAER) established in 1969 by ASCE, the Library of Congress and the National Park Service (HAER will celebrate its 30th Anniversary in 1999); additional Historic Organizations and Societies; Audio-Visual Presentations; The History of Civil Engineering: A Bibliography; a Civil Engineering Reprint Series; Special History Publications; ASCE Oral History Form; and the important ASCE Historic Civil Engineering Landmark Form. The ASCE Transactions centennial volume (Vol. CT) 1952, was devoted almost exclusively to a historical accounting of American civil engineering. Several engineering history papers are printed in the ASCE Journal: Professional Issues in Engineering Education and Practice.

Summary

The Guide to ASCE History and Heritage Programs is an educational, useful compilation of national, international civil engineering landmarks, programs, publications and resources; ASCE oral history summaries; and forms for nominating landmarks. The recently revised July 1998 "GUIDE TO ASCE HISTORY AND HERITAGE PROGRAMS" can be obtained from ASCE, 1015-15th Street, N.W., Suite 600, Washington, D.C. 20005-2605, 202-789-2200, http://www.asce.org.

HERITAGE, ETHICS AND PROFESSIONALISM IN CIVIL ENGINEERING EDUCATION

Augustine J. Fredrich, P.E.[1]

Many of today's engineering students begin their pursuit of a career in engineering with perceptions of the profession that bear little resemblance to reality. Their ideas about careers in engineering often come from teachers, counselors, relatives or friends who themselves know virtually nothing about the engineering profession, little about what it takes to become an engineer, and even less about what engineers actually do. Students who successfully complete the requirements for an undergraduate degree in engineering learn a lot about what it takes to become an engineer, a little about what engineers actually do, and almost nothing about the engineering profession!

Students who begin the study of engineering without realistic information about the engineering profession often become attrition statistics. They lack the motivation needed to endure preparatory coursework that is difficult and demanding, but offers almost nothing to satisfy their desire to learn something about engineering. Even if they manage to survive the preparatory work, they will not have the background they need to make intelligent decisions about elective courses or about areas of specialization unless something--usually something external to their formal engineering education--occurs to improve their understanding of the profession.

This paper describes a freshman-level introductory course designed to address these problems for students interested in the civil engineering profession. It uses lectures and supporting reading assignments and audio-visual materials to provide students with information on the history and heritage of the profession. Additional reading, writing and speaking assignments give students further insight into what civil engineers actually do. Students also learn about the diversity inherent in the civil engineering career field; about the diversity in civil engineering employers; about the differences in interest, aptitude and preparation for the planning, design, construction, operations and maintenance functions in civil engineering; about the various technical specialty areas within the profession; about professional registration and professional ethics; about professional societies, interest groups and trade associations; and about the meaning of professionalism.

Students find that learning about the history, heritage, scope and significance of American civil engineering becomes a fascinating exploration of their professional roots, a search for professional ancestors whose lives turn out to be as exciting and, in some ways, as familiar as the lives of American heroes they have read about since they were young children. Again and again they find civil engineers and civil engineering projects playing key roles in familiar historic stories--the Revolutionary

---

[1]. Chairman, Engineering Technology Department, University of Southern Indiana, 8600 University Boulevard, Evansville, IN 47712

War, the exploration of western frontiers, the steamboat age, the building of the Transcontinental Railroad, the Civil War, the Industrial Revolution, World Wars I and II, the Great Depression, and manned exploration of space. Civil engineers were there. Quietly doing their jobs. The unsung heroes who earned for today's civil engineers the right to call ourselves professionals.

The story of American civil engineering, they discover, is not one that is easily told. The typical engineering approach degenerates to recounting one technological achievement after another, smothering the accomplishments of engineers in tedious descriptions of their work and its technical significance. The typical historical approach shoves the civil engineer into the shadows of more flamboyant politicians, financiers and military leaders who, often by the nature of their position more than by the significance of their achievements, have captured the fancy of historians, biographers and novelists. Neither of these approaches to telling the story of civil engineering reveals the human dimensions of the men and women who practiced the profession; neither approach captures the spirit of service and the desire for tangible achievement that underlie much of what civil engineer do, and therefore, neither approach does justice to them or their accomplishments.

Civil engineering as we know it today is the cumulative work of a community of men and women with an identifiable tradition, passionately committed over a long period of time to the pursuit of a value--the creation of a built environment to serve the needs of their fellow men. In a very concrete sense, civil engineering is the heroic story of these people, bound together by a common sense of service, responding to the constantly changing needs of a constantly changing society. If civil engineering is what civil engineers actually do, and not just a collection of formulas, assumptions, theories, models and codes, then civil engineering itself is a story, a human epic of discovery and achievement as interesting and exciting as other great human adventures. In one sense each engineer's story stands on its own; in another sense, the individual stories depend on one another. None are complete until all are told. Together they form a tradition, a heritage of narratives which explore the many facets of the civil engineering profession.

The 17th century French poet and critic Nicholas Boileau wrote: "If your descent is from heroic sires, show in your lives remnants of their fires." That's a dramatic but extremely lucid way of expressing the idea that each of us bears some responsibility--first for understanding the story written through the accomplishments of our professional ancestors and then for using that understanding to keep the story alive in our professional lives.

The text used in this course, **"Sons of Martha,"** was designed by the author and published by the American Society of Civil Engineers (ASCE) to fill the very real need for easy access to stories about the civil engineering profession and its contributions to modern American society--stories which deepen one's understanding of the profession. It is a collection of writings about civil engineers and civil engineering, drawn from a very wide variety of sources which otherwise might not be readily available. The writings were chosen with emphases upon: 1) selections which reveal something interesting about civil engineers or the civil engineering profession; and 2) selections in which the quality of the writing lends itself to the difficult task of capturing the interest of readers with varying levels of understanding of the profession.

The anthology is divided into four sections: the first section contains a selection of readings which provide insights into the history and heritage of the civil engineering profession--a series of glimpses at how the civil engineering profession progressed from its crudest beginnings to the day when civil engineers ply their profession in outer space; the second section consists of a collection of biographic sketches of civil engineers--some relatively well known and some rather obscure, but all interesting human beings; the third section contains descriptions of civil engineering projects of a variety of types, carefully chosen to demonstrate the many ways that civil engineering has contributed and continues to contribute to helping our nation and its people realize their potential; the final set of readings deals with professional and ethical issues which are or ought to be matters of concern for students and practicing civil engineers alike.

Readings from the text are augmented by a series of fourteen audio-visual presentations selected to complement the reading assignments. The attached syllabus indicates the relationship between the lectures, reading assignments and audio-visual materials. Several of the supporting videos are available from ASCE; several were originally part of "The American Experience" television series broadcast by the Public Broadcasting System, and several come from commercial sources. The programs available from ASCE can be borrowed; the others must be purchased or rented, but they are well worth the investment because they are extremely well done.

In addition to these assignments, students in this course are required to keep a journal in which they must write at least three times each week. They are encouraged to use these journal entries to explore their thoughts and feelings about the civil engineering profession. Each student also prepares a biographic sketch of an American civil engineer and delivers an oral presentation on that civil engineer to the class. Civil engineers for this assignment are chosen by the instructor to insure that reference materials needed to complete the assignment are available and to minimize conflicts with other biographic materials employed in the course.

In addition to studying the ASCE Code of Ethics, the students view three ethics case studies and write brief essays on two of them. These assignments are structured in a way that encourages students to explore their feelings about ethics in general as well as about ethical conflicts that can arise in the practice of civil engineering.

Finally, students are required to complete a term paper on a civil engineering project or on some aspect of a civil engineering project. For this assignment they are free to choose what they wish to investigate, subject to the instructor's approval only to insure that each student chooses a subject that is suitable (within the limitations imposed by available library resources) and does not duplicate a topic chosen by another student.

The number of students in this class each fall ranges from ten or twelve to about twenty. Most students are initially astounded at the amount of reading and writing expected in the course; however, in the three years that the course has been offered in its present format only two students have dropped it. Journals are collected, read and responded to twice during the semester. Every student receives an assessment in writing of every writing assignment and a written response to the journal each time it is submitted for review. These activities are intended to impress on the students the importance of reading and writing in an engineering career.

It is always difficult to assess just how successfully a course fulfills its purpose when there are limited objective measurements available for such assessments. In the case of this course, however, an indication that the course achieves what it is intended to achieve is revealed in the following typical closing entries from the journals of three students in last fall's class:

> *"I would like to take this opportunity to make some comments about my thoughts on what I've learned in Intro to Civil Engineering. First of all, I was grateful for the constructive criticism of my writing skills. By drawing my attention to this particular area, you increased my awareness of the writing skills necessary at the college level.*
>
> *"Secondly, the Code of Ethics was a very important part of the class. I now fully understand and support the Code. I consider it an honor to be part of the American Society of Civil Engineers. I learned a lot about different aspects of civil engineering through the lectures, research and reading assignments."*

> *"The reading on the changing of the direction of that river was really interesting. I didn't know that that could be done. It still seems impossible. But I've learned a lot in this class. I've learned what a civil engineer does and how important it is to learn to write better. I've learned about planning, design, construction, operations and maintenance. It's been a good class."*

> *"Ethics. Career options. Professional responsibility. The Roeblings. Frank Crowe. Robert Moses. Gustav Eiffel. Corruption. James Eads. Joseph Strauss. The U.S. Military Academy. ASCE. Michelangelo. Gallopin' Gertie. Theodore Cooper. Failures. Successes. The Golden Gate Bridge. Water Resources. The Environment. Wetlands.*
>
> *"These are all things that I knew little or nothing about when I entered this course. I've learned a great deal about civil engineers and some of the things that they do. I hope one day to join their ranks.*
>
> *"Thank you, **"Sons of Martha"** for making my life a little more pleasant."*

These comments suggest that students enjoy learning about their professional heritage and that they will be positively motivated by this new knowledge. That alone is probably enough to justify the course, but the most important measure of the value of the course cannot be discerned, because it is felt rather than seen. It is the kindling of the "remnants of their fires" that instills in them the kinship with their heroic civil engineering sires.

CET 101 - INTRODUCTION TO CIVIL ENGINEERING TECHNOLOGY

SYLLABUS FALL, 1991

Text: ***Sons of Martha;*** Augustine J. Fredrich; American Society of Civil Engineers; 1989

Date	Topic	Assignments
Aug 26	Course Introduction	
Aug 28	Civil Engineering Careers	Handout & Pgs. 3-6 & 545-547
Aug 30	The Civil Engineering Profession	Pgs. 63-64, 587-594 & 91-104
Sep 2	LABOR DAY - NO CLASS	
Sep 4	Civil Engineering Through the Ages	Pgs. 7-62
Sep 6	Civil Engr Careers: Functions & Employers	ASSIGNMENT #1 DUE
Sep 9	Career Overview - Professor Naghdi	Pgs. 237-250
Sep 11	"To Engineer is Human"	Pgs. 411-416, 469-484 & 533-544
Sep 13	Professional & Technical Associations	
Sep 16	The American Society of Civil Engineers	Pgs. 65-74
Sep 18	ASCE Professional Activities	Handouts
Sep 20	ASCE Technical Activities	Handouts
Sep 23	ASCE Publications	Handout - *Civil Engineering* Magazine
Sep 25	Review	
Sep 27	EXAMINATION #1	ASSIGNMENT #2 DUE
Sep 30	Review & Professional Registration	
Oct 2	ASCE Code of Ethics - History	Handout - ASCE Code of Ethics
Oct 4	ASCE Code of Ethics - Principles & Canons	
Oct 7	ASCE Code of Ethics - Practice Guides	Pgs. 455-468
Oct 9	"The Truesteel Affair"	Pgs. 485-508 & 549-558
Oct 11	"Ethics on Trial" - Part 1	Pgs. 509-532 JOURNALS DUE
Oct 14	"Ethics on Trial" - Part 2	ASSIGNMENT #3 DUE
Oct 16	"Gustav Eiffel: Man Behind the Lady"	
Oct 18	"Gilbane Gold" - NSPE Ethics Case	Pgs. 567-584
Oct 21	"The Brooklyn Bridge"	Pgs. 293-302
Oct 23	Careers in Civil Engineering - Planning	Pgs. 255-264 ASSG. #4 DUE
Oct 25	"The World That Moses Built"	Pgs. 83-90, 417-428 & 559-566
Oct 28	Careers in Civil Engineering - Design	Pgs. 193-210
Oct 30	Careers in Civil Engrg - Construction	Pgs. 211-236 & 379-380
Nov 1	"Hoover Dam"	Pgs. 381-390
Nov 4	Careers in Civil Engr - Operations & Maintenance	
Nov 6	Review & Discussion of Career Options	Pgs. 445-454
Nov 8	EXAMINATION #2	
Nov 11	"The Roeblings - Civil Engineers"	Pgs. 137-156
Nov 13	Geotechnical & Structural Engineering	Pgs. 75-82, 307-316 & 391-402
Nov 15	"Bridges in History"	Pgs. 109-120, 157-176 & 279-286

Date	Topic	Assignments
Nov 18	Environmental & Water Resources Engr	Pgs. 251-254, 265-274 & 317-322
Nov 20	"Water Wars"	Pgs. 303-306 & 351-378
Nov 22	Transportation and Materials Engineering	Pgs. 429-438
Nov 25	"A Man, A Plan and A Canal"	Pgs. 177-192 & 323-350
Nov 27-29	THANKSGIVING RECESS - NO CLASSES	JOURNALS DUE
Dec 2	"Building the Golden Gate Bridge"	Pgs. 403-410
Dec 4	"The Iron Road"	Pgs. 121-136 & 287-292
Dec 6	Review	ASSIGNMENT #5 DUE

THE NATIONAL HISTORIC LANDMARKS SURVEY

Robie S. Lange[1]

Abstract

The National Historic Landmarks Survey was established to recognize the nation's most significant historic properties, and encourage their preservation. One area of history under ongoing study is the field of engineering. Similar to the American Society for Civil Engineers' landmarks initiative in some respects, and dissimilar in others, this federal program shares the goal of enhancing the nation's appreciation of the history of engineering, and seeks to maximize its progress through greater interaction with ASCE members, chapters, and committees.

Paper

The federal government conducts two programs to recognize historic properties--the National Register of Historic Places and the National Historic Landmarks (NHL) Survey. The NHL Survey is the older of the two, authorized by the 1935 Historic Sites Act. By this act, Congress directed the federal government (through the National Park Service) to identify and recognize properties possessing <u>national</u> significance. While an original purpose of this program was to identify potential national parks, its practical and much broader impact was to encourage the preservation of the nation's most historic properties by non-federal parties. The second federal recognition program, the National Register, was established in 1966 to identify and recognize properties primarily of <u>state and local</u> significance.

These programs have much in common, and in fact, recently were reorganized to operate side by side within a single division in the National Park Service's Washington

---

[1]Historian, National Historic Landmarks Survey, National Park Service, 1849 C Street, NW, Washington, DC 20240

office. Aside from the critical differences associated with recognizing properties possessing different levels of historic significance, a few other important distinctions between the two programs bear noting. One way to begin to appreciate the different focuses of these programs is in knowing that the National Register has recognized more than 65,000 properties, and the NHL program a more modest 2,200. This comparison not only reveals that more people have been preparing National Register nominations than NHL nominations, but that NHLs are harder to come by. There are hundreds of additional potential NHLs out there, but an NHL nomination requires more information than National Register nominations to allow an informed evaluation. It is one thing to claim that a particular water pumping station, for example, is eligible for the National Register because it was important in the late 19th century development of a major city. A successful NHL nomination, however, must demonstrate an understanding not only of the specific history of that pumping station, but of the national context of the history of water supply to establish why this particular pumping station should be selected for NHL designation.

National Register designations are made by a National Park Service program manager known as the Keeper of the National Register, in collaboration with each state's State Historic Preservation Officer. NHL designation authority resides solely with the Secretary of the Interior.

Another important distinction between the requirements of the two federal recognition programs relates to the level of historical integrity required for eligibility. While National Register criteria require properties to possess historical integrity, NHL criteria further qualify this standard to require eligible properties to possess a **high degree** of historical integrity. Without delving too deeply into the fine points of how cultural resource experts apply this test, the concept of historical integrity largely relates to whether the historic resource still maintains the physical attributes it possessed during its historic period. It is especially critical for physical attributes that most importantly relate to the property's national significance to possess a high degree of integrity. For example, a dam believed to be nationally significant for its association with rural electrification would likely be found not to possess sufficient historical integrity if much of its powerhouse equipment had been replaced. One the other hand, a dam nominated for its association with an important advancement in concrete construction methods might be found to possess the required degree of integrity if the nearby powerhouse has been modified yet the dam itself continues to reflect the original design and construction. Admittedly, the measurement of this concept incorporates subjective considerations, yet guidelines have been developed to assist such evaluations. While two people may honestly disagree on where to draw this line, the important point to understand is that NHLs are expected to meet a very high standard in this regard.

Many of the properties already found to meet NHL criteria are those most of us have heard of, such as the Brooklyn Bridge, Hoover Dam, Springfield Armory, and Henry Ford's Highland Park Automotive Factory. Many other NHLs, however, are less well known or appreciated by the general public. New York and New Jersey's Holland Tunnel is a good example. By the 1920s, metropolitan areas were rapidly expanding beyond the rivers which confined many of them, and automobiles had become a major agent in that development. In some cities, a bridge crossing was not technologically or economically feasible, whereas a tunnel might be more suitable provided that adequate ventilation could be incorporated to allow automobile drivers sufficient air quality to operate their vehicles without being impaired by exhaust fumes. The design of the Holland Tunnel required not only the usual engineering studies, but an understanding of the nature of auto emissions and their impact on drivers, and how to produce an effective and efficient means to maintain carbon monoxide at safe levels. The solution developed for the Holland Tunnel has been a basic part of virtually all subsequent river automotive tunnels. Key aspects of this ventilation system are visible while driving through these tunnels--openings for forced-in fresh air at the tunnel's curb level, and out-take openings for the warmer, rising fouled air in the false ceiling overhead. Even though Holland Tunnel has the distinction of being recognized by ASCE and ASME as an engineering landmark, it was with its NHL designation that New York Times articles and national radio stories brought this history to not only thousands of the tunnel's daily commuters, but to a nationwide audience.

The regulations that guide the NHL designation process identify several types of properties that ordinarily are not eligible for designation. These include cemeteries, birthplaces and graves of historical figures, religious properties, structures that have been moved, reconstructed properties, and properties that have achieved their national significance within the past 50 years. These same regulations outline certain exceptions to these rules by which such properties might be eligible for NHL designation. Most of these exceptions to the rules are applicable when it can be justified that the nominated property's historical association is of the highest order even in relation to the already high standards for NHL designation. Examples of this include properties strongly associate with pivotal events in the past 50 years such as Little Rock High School, the scene of the school desegregation clash of 1957, the site of President Kennedy's 1963 assassination (Dallas' Dealy Plaza), and the Apollo Mission Control Center.

Ideally, NHL nominations emanate from theme studies which examine an aspect of U.S. history such as the Civil War, labor history, the underground railroad, modern architecture, the automobile, or tunnels. The main advantages of this approach is that the necessary historic context is developed to allow a well informed judgement as to whether the best properties associated with a common theme have been singled out for nomination. Despite this important advantage, individual properties are frequently nominated without the benefit of a theme study. In those cases, the individual nomination bears the burden of establishing the broader historic context, and the nominated property's national significance and relative merit.

The great majority of NHLs of likely interest to ASCE members were nominated under the overall theme titled "Technology (Engineering and Invention)." The more specific subtheme headings under which properties have been designated include: surveying, transportation, tools and machines, military fortifications and weapons, extraction and conversion of industrial raw materials, construction, space exploration, and water, sewerage and sanitation.

To allow a better idea of what properties might meet NHL criteria, here are several examples of designated NHLs in each of the subthemes associated with the history of engineering. An NHL designated for its nationally significant association with surveying is the Beginning Point of the U.S. Public Land Survey, on the Ohio and Pennsylvania border. This NHL represents the administration and subdivsion of the old Northwest Territory beginning in 1785. The subtheme of transportation is heavily represented with NHLs reflecting the span of U.S. history. These include the 1783 Boston Lighthouse, representative of the importance of maritime transportation; the 1813 Casselman's Bridge along the western Maryland portion of the National Road; and Detroit's 1917 Lincoln Motor Company Plant. NHLs associated with energy utilization range from the early 19th century system of locks and canals used to power Lowell, Massachusetts' textile mills, to New Mexico's Los Alamos Scientific Laboratory. A good example under the subtheme of tools and machines is Windsor, Vermont's mid-19th-century Robbins and Lawrence Armory and Machine Shop. NHLs designated for their association with military engineering include Norfolk Naval Shipyard's 1827 Dry Dock One, and the first nuclear submarine, U.S.S. Nautilus, now located in Groton, Connecticut. The extraction and conversion of industrial raw materials are represented by Michigan's Calumet and Quincy Mining Company historic districts, and Birmingham, Alabama's Sloss Blast Furnace. NHLs under the subtheme of industrial production processes include Illinois' Rock Island Arsenal, and southwestern Pennsylvania's Cambria Iron Company. The subtheme of construction contains many entries, such as New York's Old Croton Aqueduct, completed in 1842; and the St. Clair Tunnel completed in 1891 at Port Huron, Michigan, as the first subaqueous tunnel. Under information processing, transmission and recording, NHLs include the Rochester, New York, home of George Eastman, a leader in the development of photography as a popular medium; and the Newton, Massachusetts, residence of Reginald Fessenden, a pioneering inventor in the field of radio. The Earth and space exploration subtheme includes the site of Robert Goddard's 1926 launch of the world's first liquid-propelled rocket in Auburn, Massachusetts; and a few properties associated with the Apollo moon mission project of the 1960s, such as Huntsville, Alabama's Dynamic Test Stand for the Saturn V rocket. Finally, water, sewerage, and sanitation are represented by Philadelphia's 1812 Fairmount Water Works, San Diego's 1817 Old Mission Dam, and Shreveport, Louisiana's late-19th-century Waterworks Pumping Station.

Again, there are many potential NHLs associated with these themes which have not yet been nominated. Engineering properties currently under review for NHL nomination include San Francisco's Golden Gate Bridge, Montana's Fort Peck Dam, the last surviving Bollman Truss Bridge in Savage, Maryland, and Houston's Northside sewage treatment plant, a pioneering activated sludge treatment facility.

Anyone can prepare and submit an NHL nomination. Interested parties should contact us before initiating an NHL nomination. We respond to initial inquiries with a preliminary evaluation of NHL eligibility, which often includes a list of research questions tailored to the specific property that will help lead to an informed evaluation of national significance. We can help in a number of ways in offering guidance about details of preparing nominations. This initial evaluation is usually very helpful since the great majority of nomination preparers have never undertaken this task before, and once their property is considered, may never prepare a nomination on another property.

After a nomination is completed it is ready for the formal nomination process. The nomination is included on the agenda of the next meeting of the National Park System Advisory Board, which meets twice a year. The Advisory Board is composed of citizens and scholars appointed by the Director of the National Park Service. Since NHL regulations require that property owners, elected officials, and others have at least 60 days written notice before a property is considered for NHL designation, the nomination must be completed more than two months before the next scheduled Advisory Board meeting. Shortly before the Advisory Board meets to consider these nominations, a committee of the Board (the National Landmarks Committee) convenes in a public meeting to hear brief presentations on each nomination, receive comments by owners or other interested parties, discuss the merits of the nomination, and then vote whether to recommend NHL designation to the Board. When the full Advisory Board meets to take up NHL nominations along with its other National Park Service business, it receives the report of its Landmarks Committee, entertains any public comments, and votes whether to concur or disagree with the committee's recommendations. The favorable recommendations are then forwarded to the Secretary of the Interior for a final decision about designation.

Owners of private properties can object to, and thus prevent, NHL designation. Owners of designated NHLs are under no obligation to open their properties to the public. Private property owners can do anything they wish with their properties, provided that no Federal license, permit, or funding is involved. When Federal agencies are involved in a project that affects an NHL, the Advisory Council on Historic Preservation must have the opportunity to comment on the project and its effects on the property. The most visible result of NHL designation is that the property may receive a bronze plaque for public display attesting to the landmark's national significance. Federal historic preservation funds and investment tax credits for rehabilitation may be available to owners of designated NHLs, as is technical preservation advice from National Park Service professionals. Many individuals and organizations seek NHL designation because this status may aid them in competing for funding or other preferences from Federal or non-Federal sources.

Currently only one-third of ASCE landmarks are also NHLs. The remaining two-thirds (more than 100 properties) have either been judged not to meet NHL requirements of national significance and high integrity, have not been formally nominated for NHL consideration, or have not been nominated or designated because of owner objection. At the same time, there are many National Park Service NHLs that have not been considered for ASCE recognition, yet might meet its requirements. We would welcome anyone who is interested in reviewing our files to support his own preparation of new ASCE nominations. Perhaps ASCE would consider designating the few historic properties that are not NHLs simply due to private owner objection to federal designation. In one such instance, we nominated the 1910 Detroit River Railroad Tunnel as the first full-size application of the trench and tube method of subaqueous tunneling. Devoid of even the modest implications of federal landmark designation, perhaps ASCE designation would be more acceptable to the tunnel's owner. Interested ASCE members are encouraged to look into this and I invite them to use the National Park Service-prepared nomination in any way they see fit. We hope to identify ways to consider more current and future ASCE landmarks for NHL nomination, and to share future NHL-sponsored nominations with ASCE for its consideration. In broader terms, we hope to work with your History and Heritage committee to identify additional ways to maximize our mutual goal of enhancing the public's appreciation of the contributions engineers have made to our nation's history.

The Civil Engineer at the Smithsonian Institution

William E. Worthington, Jr.[1]

Abstract

Civil Engineers should feel welcome at the Smithsonian Institution, as it is one of the few museums in the world that deals with their work in a substantive way. The subject of the built environment is documented, studied, and interpreted in collections both on permanent display and in storage, and in extensive archival holdings. In these collections we are preserving the history of the American civil engineering profession. It is an engaging, challenging, and always interesting undertaking.

Introduction

The rationale that guides the collecting of engineering history differs little from that used by social and cultural historians. Our mission is to collect, study and interpret the material remains of the past. In carrying out this work we have amassed both large and small three-dimensional objects, including pieces of structures, patent models, and samples. They are complimented by large holdings of two-dimensional artifacts which include historic photographs, paintings, prints, films, and engineering drawings that further document engineer's activities. But, the work of the civil engineer presents the historian with peculiar challenges. As the engineering collection has evolved, it has been necessary to devise different collecting strategies. Unlike items used in the home or at work, the things created by engineers are most often built into the landscape. If these constructions can be removed at all, they are usually far too large to collect in the traditional sense. Our approach has been somewhat roundabout and in many cases it must be through objects and documents that are only secondary to the actual project under consideration. This has led us to collect representative examples--bits and pieces if you will--of and about engineering works. Thus, for example, in lieu New York City's George Washington suspension bridge we have the full-size, 16-ton, cable sample laid up prior to the time the bridge was

---

[1]Museum Specialist, Engineering Collections, Room-5014, MRC-629, National Museum of American History, Washington, D.C. 20560

actually built. The collapse in 1907 of the first Quebec cantilever bridge is documented by a sheared off rivet head and a broken hex nut taken from the wreckage. Our understanding of these and other artifacts is further enhanced by extensive reference and photo files where we have gathered any and all information on the various subjects in our charge.

Civil Engineering

The Smithsonian's effort to document the history of civil engineering is relatively recent. Although the institution was chartered in 1846, it was not until 1942 that the first historically significant civil engineering object was acquired. At that time, the Reading Railroad donated a composite cast- and wrought- iron bridge truss designed by Richard Osborne in 1845. It was one of a set of three that made up America's first iron railroad bridge. The bridge was retired to a public park in 1901 when the increased weight and speed of rail traffic rendered it obsolete.

Although we have one part of that first bridge, vast numbers of 19th century bridges have gone completely unrecorded. There is little to be collected about them as the bridges themselves are long gone and there are few engineering drawings extant earlier than the 1870s. Recently, we were extremely fortunate to acquire an album of drawings and photos of the Chicago, Burlington & Quincy railroad bridge built across the Mississippi River at Burlington, Iowa, during 1867 and 1868.

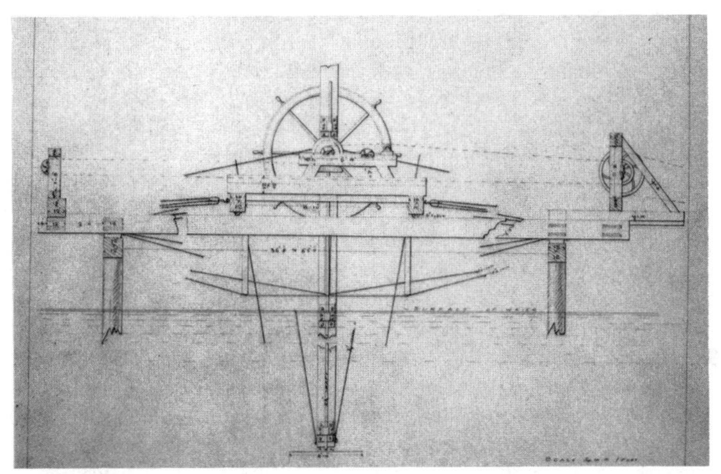

Figure 1. Among the equipment designed especially for the Chicago, Burlington & Quincy Bridge project was this underwater saw.

This hand-drawn album was the property of project's resident engineer Charles H. Hudson. By the time the bridge was under construction, Harvard graduate Hudson was well versed in the field having worked for numerous railroads in both the east and mid-west. The album is a rare find because of its age and attention to detail. But most importantly its pages contain a complete set of dimensioned drawings for the bridge and the equipment used to erect it. A group of photographs at the back of the album compliment the drawings and show the work in progress. Of particular significance are the stress diagrams for the trusses. We are not sure why this document was created or whether it is in the hand of Hudson, but it is unique in its completeness.

In 1996, we received a gift of what probably is the most important record of American civil engineering history to be found in this or any other museum. It is a collection of over 30,000 engineering drawings from the Lockwood Greene company. Lockwood Greene is perhaps the nation's oldest active engineering firm as it can trace it roots back to 1832. New Englander David Whitman began a textile machinery repair business during Andrew Jackson's first administration. Business was so good or, put another way, mills were so bad that Whitman's operation grew and prospered to the point where he offered mill design services as well. Design continued as the primary focus throughout the century. Principals changed, and in 1882, reflecting the names of the current partners Amos Lockwood and Stephen Greene, the firm became Lockwood, Greene & Company, Mill Engineers.

Originally located in Boston, the firm identified itself as engineers and architects. Living up to the stereotype of the thrifty New Englander, the company preserved nearly its entire output of drawings from the time of its formation in the early 1880s. In breadth and scope, the collection is an unrivaled record of American engineering practice over the last 100 plus years and it stands as an unparalleled account of our country's industry, business, and the evolution of technology. A number of its riches have already come to light and there is no doubt that many more will be found. The story of the collection is a good example of what our work involves and what can be learned from long silent engineers.

The drawings were stored for decades in two vaults in a Cambridge, Massachusetts, warehouse. Occasionally there was a need for some of the drawings, as the company maintained relationships with some longtime clients. But the bulk of the collection lay unused for many years. Recognizing the historical importance of the collection, Lockwood Greene determined to insure its long-term preservation by getting it into the hands of a museum. The drawings were offered to and accepted by the Smithsonian Institution.

Reacting to the growth in the construction of textile mills in the South, Lockwood Greene opened one of its first branch office in Greenville, South Carolina, in the late 1890s. There is no question that the engineers who directed Lockwood Greene in the late 19th century were pragmatic. Where the textile industry had been almost an exclusive client for most of the century, changes in that industry after 1900 led the company to accept increasing amounts of work from other quarters. This along with improvements in business conditions led to other branch offices in a number of cities, most notably Chicago.

Thanks to the breadth of the work they undertook, we are left with an impressive diagrammatic history. Along with drawings for all sorts of factories, including those making shoes, pianos, automobiles, furniture, and wire rope, there are those for highways, airports, radio stations, private residences, banks, hotels, department stores, textile mills, and American industry in general. The importance of these documents is driven home when one realizes that they are the only surviving record of many structures. Where buildings were altered or completely removed, and industries disappeared, the drawings can provide information that is found nowhere else.

For example, operating speeds for steam- and waterpowered equipment are not well documented, and in many instances can be only roughly estimated. Such information was likely to have been transmitted from one operative to another by word of mouth, or as sketchy notes. Methods existed for calculating speeds for individual machines, but determining those for an entire factory would be difficult. Here, for the first time, we find listed pulley sizes, the location of shafting and machines, and the actual operating speeds for everything from steam engines, waterwheels, and rope drives to the factory machines themselves.

A 1903 tracing of lighting plans for the Brandon Mills at Greenville, South Carolina, that specifies the locations of ceiling-mounted lamp fixtures, actually provides information about working conditions in the early electric age. Compared to what we have come to expect as a reasonable level of illumination in the late 20th century, textile mills dating from the turn of the century were dim places to say the least. In some areas ceiling outlets are shown spaced 16 feet apart. More amazing is that the drawing calls for the use of bulbs of only 50 watts. Such a vignette of factory life is not readily found elsewhere.

During the 19th century, water powered installations were a large segment of Lockwood Greene's design work and we find countless drawings related to hydraulics. While there are drawings for races and dams, details of the wheels (turbines) usually supplied by a recognized manufacturer generally are not to be found. In some situations specialists were brought in and their drawings are found among those generated by Lockwood Greene. One of the more interesting finds was a rare blueprint for an Ambursen dam. This early 20th century product of the Ambursen Hydraulic Construction Co. of Boston is known but not well documented. Here is a drawing of one of their patented dams.

For many years Lockwood Greene mills were quite traditional in their configuration. Power from waterwheels or turbines was transmitted by leather belts or endless ropes, pulleys, and line shafting to a factory full of machines. All this changed with the design and construction of the Columbia Mills in Columbia, South Carolina, in 1893. For the first time, power came to a factory through wires and machines were run by electric motors. As a result, factories no longer needed to be situated at water's edge. Much depended on the success of this first installation. Had the project's electrical component not functioned properly, Lockwood Greene would have been in dire straits, for, needless to say, the mill building could not easily have been moved. The firm met with similar success with the Soule Mill in New Bedford in 1901, one of New England's earliest electrified textile mills.

When Americans took to the roads in the automobile in the early 20th century,

Lockwood Greene played a modest role in their travel plans. The Lincoln Highway, conceived and begun in 1913, was the nation's first major east-west motor route. The project got off to a good start in some areas, particularly around larger cities that lay in its path, but it languished in many places for lack of funds. In the early 1920s in an effort to rekindle flagging public interest, promoters hit on the idea of an "ideal section." This one-mile stretch designed with perfect alignment and all the amenities would be built in Lake County, Indiana, near the route's midpoint. Lockwood Greene engineers produced a set of drawings for a substantial concrete roadway, improved lighting, signage, and drainage.

Not only was the roadway to be the best possible, the facilities for the comfort of travelers were to be the most modern. Lockwood Greene's detailed drawings for the ideal autocamp give us a sense of what long-distance auto travelers faced in the 1920s. In a carefully set-out plan of tent sites, fireplaces, and trees, the amount of space allotted for each vehicle is duly noted. A service station and convenience store completed the facility. It is interesting to note that a separate structure was planned solely for the greasing of automobiles. This is telling information about the challenges motorists dealt with in terms of both road conditions and the machines themselves. The campsite was never built, however. There was such a delay between the time of design and construction that motorists tastes had changed to the point where they preferred the less rustic accommodations of motor courts and tourist cabins.

Paper drawings dated 1929 for an airport at Woonsocket, Rhode Island, provide us with a glimpse of early commercial aviation. Along with several configurations for the landing strips, there are hangars, machine shops, waiting rooms, and an eating area. At the end of one strip is a parking lot and spectator bleachers. It was only two years since Charles Lindbergh had landed in Paris, and nationwide air races and barnstorming were still popular. More importantly, savvy air liner operators and promoters understood that if air travel was to become an industry and an integral part of American life then the public had to become "air minded." Engineers included these and other features in order to attract anyone who cared to observe the novelty of flight first hand.

Like many American businesses, the textile industry created company towns in which to house its workers. The collection documents housing for a number of mills. As might be expected, managers' residences tended to be larger and more finished. Some plans include details of railings, sills, doors, and even the location of dining-room tables. Worker housing was smaller and tended to be less detailed. Some town schemes were rounded out with boardinghouses, and multipurpose school-church and library buildings.

In 1917 the directors of Marshall Field & Co., the well-known Chicago department store, sought to develop a modern version of a company town around a Field-owned textile mill in Virginia. There would be no squalid shanties sitting cheek by jowl. This was to be the Community of Fieldale. A series of site plans show the location of housing along tree-lined streets and avenues. There were playing fields and recreation areas, schools, a theater, and a hospital. The mill would be reached by bridges across the streams and valleys that intersected the site. Characteristics of this never-built town bear a striking similarity to what we accept as typical in late-20th-century planned communities.

Among the drawings is evidence that documents advances in structural technology. If the Maverick mill in East Boston built in 1909 was not the first reinforced textile mill it was one of the earliest. In drawings for the George N. Pierce Company, builder of Pierce Arrow automobiles, Lockwood Greene made modifications and additions to Albert Kahn's first reinforced-concrete automobile factory. Blueprints of the Dixie textile mill in LaGrange, Georgia, are an interesting record of what turned out to be an ephemeral building process, patented by Charles A. Praray in 1894. He was a principal of the rival Boston mill engineering firm of C.A. M. Praray & Co. Lockwood Greene was hired to make improvements to the building and its engineers drew plans of the existing structure that show the footprint of Praray's zigzagging steel-and-glass curtain walls. This strange pattern was especially useful in textile mills where walls made primarily of glass would flood an interior with natural light. Although the mill building still stands, enlargements over the years have literally encompassed and obliterated Praray's geometric design.

The march of technology turns up in some not-so-obvious ways. A 1920s pencil sketch intended to show the facade of a now-unidentified hotel or office building reveals several unrelated yet fascinating items. The drawing comes to life in these details. The building may have housed a radio studio. On its roof are two towers that support the wires of the type of aerial that was used for broadcasting during radio's infancy. Below it, directly on the roof, is a signboard proclaiming "Tunney Wins!" That information, a reference to one of the 1926-1928 boxing matches in which Gene Tunney defeated Jack Dempsey for the heavyweight boxing title, may have just been received over the airwaves.

Radio transmission and reception had changed to some extent by the 1940s when Lockwood Greene prepared the plans for several radio stations, including one in rural Montgomery County, Maryland. The trend was toward facilities dedicated to broadcasting and drawings were made for everything from the structural steel tower to the broadcasting studio and details of the call-letters sign. The facility was small but thoroughly documented.

The Lockwood Greene Collection eventually will be open to researchers and historians, but the process of preparing it for use is ongoing at this time. The wait will surely be worthwhile. There is no question that many more treasures will surface as we literally unroll Lockwood Greene's engineering legacy to the nation.

Photos are particularly useful in revealing engineering practices and the manner in which projects evolved. For instance, a photo album that traces construction of one part of the Boston elevated railway in 1902 helps us to understand how large works were carried out prior to the advent of mobile, heavy power machinery. It also shows us that little has changed in the way in which urban life has had to deal with ongoing engineering projects.

Lest I paint a one sided picture of our successes, there are also examples of our national patrimony that have eluded us. The systems that complete and make a structure usable are important artifacts. Aspects of heating, lighting, ventilating, hoisting and conveying are all of interest to us, and we have examples of furnaces, fans, pumps, cooling systems, and elevators. One particularly unfortunate loss occurred in Boston. Until a few years ago one of the world's oldest escalators was still in regular service in Boston's transit system. A search of the patent office records revealed that the first escalator was patented in the 1850s. It was not translated into a useful device, however, until the 1890s as there was simply no need for it prior to that time. We learned that one of the escalators leading from South Station was installed in 1914 and its design was a direct precursor of the technology in use today. Although the moving components were quite different, the truss on which it operated was not unlike a simple bridge truss. As the city upgraded its system and old equipment was removed, the escalator was slated to be replaced. Despite its significance and a general interest in the project, we were unable to gather enough support to have it removed in such a way so that it could be preserved in the Smithsonian.

Figure 2. A 95-ton steam shovel at work on the Panama Canal. More than 100 shovels were used on the project and most were supplied by the Bucyrus Company.

The construction of the Panama Canal, one of the 20th century's first major engineering achievements, is well documented in photographs. But, there is little from the canal itself that can be collected--aside from the fact that it is still in use. One significant item that remains in our collecting plan is an example of the steam shovels used to dig sections of the canal. These along with the railroad were major factors in the American success in Panama. The shovels are well documented. In fact, one of the best known photos made during the work shows President Theodore Roosevelt sitting at the controls of one of the shovels. We know the names of the manufacturers, exactly how many were purchased, and the cost of each. Some machines were lost in slides, others were simply used up. Of those still operational after the canal was completed, several were held in reserve until the 1930s for possible canal repairs that were never needed. Records indicate that many of the machines that were sold went to South America. Over the years we have received reports on and offers of shovels supposedly used in Panama. The machines have been in such distant spots as a sand and gravel pit in Tennessee and the tundra in Alaska. Thus far, however, none of the prospective donors has been able to provide conclusive proof that their machines were used in Panama. To our benefit, official documentation provides a description of and serial number for each of the machines purchased by the Canal Commission.

Nonetheless, we have documented the work in Panama through other sorts of artifacts. Aside from the typical photo albums of engineers, post cards, and panoramic views in the collection, there are several unique specimens. When the French effort began in the early 1880s under the direction of Ferdinand de Lesseps, stock was sold to finance the project. But, as investor interest waned, the canal company turned to lotteries as fund raisers and we have a ticket issued during one of the campaigns. Despite investor belief to the contrary, part of the money raised in the lottery actually went to buy equipment with which to dig the canal. Although the machinery used by the French worked reasonably well in the marshy areas near the coast, it was ill-suited and not up to the task of the rocky terrain it faced inland. American engineers later were to remark that that ground was some of the most difficult they had ever encountered in any locale. It is no wonder then that much of the old European equipment was simply left abandoned when it ceased to operate. By the time the American's took over the project in 1904, the jungle was littered with rusting and useless machinery. Brass nameplates from several of the Belgian-made machines are one way of documenting the failed French effort.

One of the more curious objects related to the Canal is a wooden walking stick donated to the museum in 1915 by the widow of Col. David Gaillard, one of the American engineers. The stick was fashioned from a log found during the work. On the face of it this is not very exciting, but when it is revealed that the log was discovered 64 feet under ground during excavation for the lower Gatun lock we can better appreciate the conditions engineers faced. They complained that some areas were so unstable that railroad trestles had to be replaced time and again as they would simply disappear into the ground. Perhaps the same thing happened to the tree that provided the wood for the cane.

In documenting the various aspects of civil engineering we have collected artifacts that are unique as well as examples that happen to represent extremes in size. A Riehle 300,000 lbs universal testing machine fits nicely in the former category. This machine, known as the "President" model, aside from being the largest one available, was specially prepared for exhibit at the World's Columbian Exposition in Chicago in 1893. It was bought at the exposition display floor by Purdue University and spent its entire life in the school's civil engineering lab. It came to the Smithsonian in the mid-1980s when the university modernized the facility.

In the 1870s a controversy over the merits of cast iron versus steel as the best metal for the manufacture of ordnance led the U. S. Congress to appropriate money for an improved materials testing machine. This resulted in a testing machine being built in 1873 by Albert H. Emery. It continued in use until the early 1960s at the U.S. Army Arsenal at Watertown, Massachusetts. Constructed with a movable carriage and two 45 feet-long horizontal screws, it was one of the largest mechanical testing machines ever built, and it was capable of exerting compression and tension stresses of 80,000 lbs. At the time it was constructed and well into the 20th century it was considered to be the most accurate testing machine made. On the lighter side--physically, we have early examples of the tiny and delicate Baldwin-Lima-Hamilton SR-4 electric strain gages and the equipment used to take readings from them.

Conclusion

Although it is possible to cite only a few examples from the vast National Collections, it should be obvious to even the most casual observer that objects do not stand by themselves. Only a brief look beyond the surface reveals that history invariably connects objects to people, events and other objects. There is always a story in this relationship and usually it is an interesting one. These stories help to tell us why history has unfolded as it has, what has been done, and why we do things in certain ways. Ultimately objects can help guide us, if we care to take the time to look beyond the surface.

Advancing the Awareness of Civil Engineering History

Irving Sherman, MASCE[1]

Abstract

The work of advancing the awareness of civil engineering history in Southern California, both within the Los Angeles Section, ASCE and by other organizations is described.

Introduction

The ten-county area of the Los Angeles Section is not only geographically large at 61,389 square miles, but is also one of the areas fastest growing in population in the United States. Since the end of World War II in 1945 the population has grown from about three million people to over 16.6 million in 1995, half of the population of the State of California.

As a result, collective community memories tend to be short. Many of the residents came from other areas and countries, and are largely unaware of the general history of their new home, let alone of the civil engineering that made the development possible. And because such a vast proportion of the built environment is so recent, the older historic civil engineering works may be visually submerged by the newer construction.

So there is a great need to help both the general public and also the members of our own profession to become more aware of the history and heritage of civil engineering in Southern California. This paper presents an overview of the continuing efforts both within and outside of ASCE to advance the desired awareness.

----

[1]Consulting Civil and Geotechnical Engineer, 24030 Killion Street, Woodland Hills, CA 91367.

Activities within the Los Angeles Section

The primary responsibility for making civil engineering history more widely known has been assigned by the Section Board of Directors to the History and Heritage Committee. This group of dedicated volunteers with a current membership of fourteen also works closely with the History and Heritage Committees of those Branches within the Section which have organized such units.

The activities carried on by the History and Heritage Committee have included:

Recommending projects to the Section Board for designation as Historic Civil Engineering Landmarks. (There are currently 62 such designated projects within the Section area.)

Preparing memoirs of distinguished deceased Section members, for publication in the Section Newsletter and in ASCE _Transactions_.

Interviewing distinguished civil engineers and taping the interviews for the fledgling oral history program.

Preparing slide programs for presentation to various audiences.

Preparing brochures illustrating our Historic Civil Engineering Landmarks.

Making presentations on civil engineering history at Section and Branch meetings.

Organizing ceremonies for plaque presentations at newly designated Historic Civil Engineering Landmarks.

Sending photographs and texts concerning Landmark projects for publication in ASCE News and in Civil Engineering magazine.

Making a poster presentation at an ASCE National Convention.

Making recommendations to the Section Board for celebrating the Section's 75th Anniversary.

Proposing a U.S. Postage stamp to celebrate ASCE's 150th Anniversary in 2002.

The Chairman of the Section's History and Heritage Committee also currently chairs the Standing Committee on Historical Civil Engineering Landmarks of ASCE's California State Council.

Of course other units of the Section also contribute to the History and Heritage effort. The Board of Directors provides direction and funding. The Branches participate enthusiastically in helping organize Landmark plaque dedication ceremonies within their areas. The Directories Committee placed a list of Historic Civil Engineering Landmarks in the latest Membership Directory, along with map locations for each, so that interested members could more easily visit the projects. The Newsletter Committee publishes History and Heritage articles and photographs. A Task Committee including representatives of several standing committees organized the celebration of the Section's 75th Anniversary in 1988-89, and published an Anniversary Calendar which included photographs of a number of Historic Civil Engineering Landmarks. The Public Relations and Publicity Committee tries to get media coverage for our events.

Promoting Historic Awareness outside of ASCE

1. The Huntington Library.

We are most fortunate to have in Southern California a research library with a notable collection of materials on civil engineering history. The Library was endowed by Henry E. Huntington, who built the Pacific Electric Railroad, so his interest in railroads led to the Huntington Library collecting many books and manuscripts about railroads as well as other civil engineering works.

We are also very fortunate in having as a Section member Mr. Trent R. Dames, FASCE, a past Vice President of ASCE who started ASCE's History and Heritage Program with a generous endowment over 30 years ago. Mr. Dames has also endowed the "Fund for the Heritage of Civil Engineering" at the Huntington Library, which has greatly expanded the work of the Library in civil engineering history. Recently the Library showed some of its collection in an exhibit entitled "Directing Nature: The Engineering of Our World", and arranged for Mr. Eric DeLony, Head of the Historic American Engineering Record, to present the first annual Trent Dames Lecture to the public. The Library has provided space for ASCE meetings, speakers to address ASCE groups, and has kept ASCE advised of events such as the recent meeting in our area of the Society for the History of Technology. Two members of our Section's History and Heritage Committee now serve on the Advisory Board of the Fund for the Heritage of Civil Engineering.

2. Other organizations

We have had support from a number of outside organizations. Particularly noteworthy was the contribution by Southern California Edison Co., which financed publishing our 1974 brochure "Historic Civil Engineering Landmarks of Southern California".

The owners of projects which have been designated as Historic Civil Engineering Landmarks have been most cooperative in publicizing the ceremonies at which the ASCE plaques were unveiled. These owners have recently included the U.S. Marine Corps, the Metropolitan Water District of Southern California, the Los Angeles County Department of Public Works, the Southern California Edison Company, the City of Newport Beach, the U.S. Army Corps of Engineers, the Catholic Diocese of Orange, and the California 32nd District Agricultural Association.

It has been very gratifying that the Los Angeles Times (Daily circulation 1,095,007, Sunday circulation 1,385,3730), one of the major newspapers in the United States, has recently run occasional long articles about historic civil engineering works, such as the Four Level Interchange, the original hub of the Southern California freeway network. The Times' reporters who specialize in such articles telephone the ASCE History and Heritage people whom they have come to know on a first name basis, to verify information and get explanations. Other smaller newspapers also publish shorter articles about civil engineering matters from time to time, but we consider it a major accomplishment to have the Times consider that civil engineering history can be newsworthy.

Local governments have also been supportive of our efforts, even when they were not owners of our Landmark projects. Recent plaque ceremonies have had participation by local mayors, members of City Councils, and members of County Boards of Supervisors. Their presence at such events shows their awareness of the importance of civil engineering works in the development and functioning of their communities.

Summary

Many people and many organizations have contributed to advancing the awareness of civil engineering history in Southern California. Much has been accomplished, but much more remains to be done in helping educate a large public and even our own ASCE members about the accomplishments of those earlier engineers on whose shoulders we now stand.

The Inland Empire Section - ASCE and the Grand Coulee Dam National Historic Civil Engineering Landmark Ceremony

Robert B. Turner, P.E., M.ASCE[1]

Abstract

This summarizes a review of activities associated with the Grand Coulee Dam National Historic Civil Engineering Landmark Ceremony held at Coulee City, Washington, Saturday, March 21, 1998.

Introduction

The process of recognizing a local Historic Civil Engineering Landmark must follow a defined procedure. The Inland Empire Section created committees to track progress, make formal application, organize activities associated with the ceremony, and ensure success. This report describes the process involved in creating an interesting and successful effort in getting a local Historic Civil Engineering Landmark recognized as a National Landmark.

[1]City of Spokane, Transportation Department, 808 W. Spokane Falls Blvd, Spokane, WA, 99201-3314

To be successful in having a local project recognized as a National Historic Civil Engineering Landmark, it is necessary to understand the details required to make this happen and then properly divide the responsibilities and tasks.

The Inland Empire Section of ASCE did just that as it pursued recognizing Grand Coulee Dam. There were certain steps that the section followed to manage this project to a successful completion. The following points served the Inland Empire Section well in having Grand Coulee Dam recognized as a National Historical Civil Engineering Landmark.

1. Identify potential projects.

2. Develop a scope and schedule for the project.

3. Organize a committee and build a strong team.

4. Focus on the requirements and time constraints necessary to make formal application through ASCE.

5. Pursue the project with the mindset of creating a showcase for ASCE and Civil Engineering in general.

6. Select a project historian/secretary.

Point number one, ***IDENTIFY POTENTIAL PROJECTS***. The act of identifying projects, old or young, is significant for a section or branch. This effort helps to build pride, establishes a necessary historical interest and perspective, and helps sections and branches be better prepared to submit projects for recognition, whether locally or nationally such as the annual Outstanding Civil Engineering Achievement Award.

To specifically address the issue of a National Historic Civil Engineering Landmark, a section should perform an inventory of significant Civil Engineering projects in the section or branch that are 50 years old or older. You can also take suggestions from others. The Pacific Northwest Council (PNC) of ASCE, a collection of sections from Zone IV, threw down a challenge to the Inland Empire Section to recognize Grand Coulee Dam and hold the ceremony when the section next hosted the PNC council meeting.

Point number two, ***DEVELOP A SCOPE AND SCHEDULE FOR THE PROJECT***. Determine significant milestones, such as application deadlines, date at which the project will be 50 years old, and in general, allowing enough time to compile the significant amount of data necessary to support the nomination of the project.

Remember, the most important thing to consider, like any good engineer does, is just how much time it takes to get a project done!

Begin by determining when the best time would be to hold the dedication. This might be controlled by time of year, some anniversary associated with the project, or a scheduled section or branch event, such as an annual meeting.

Start as early as possible. To emphasize this, invitations sent to national elected officials were declined because their calendars were already full on the prescribed day, even though the initial invitation was made approximately 14 months in advance!

Be sure to also start early in establishing contacts with those individuals who need to be involved. The Section needed to make contacts with federal officials associated with the Dam who were very busy and were already calendared 12-18 months out. Don't forget to consider ASCE officers as we know that their schedules can become very restricted. We suggest anyone getting involved in this process should recite the mantra, Start Early, Start Early, every day as you decide to enter this adventure.

Point number three, ***ORGANIZE A COMMITTEE AND BUILD A STRONG TEAM***. We suggest that you select one person to chair this process and be a central coordinator for all the activities associated with the project. Our section initially looked for someone who was retired and history minded. We were not able to locate any willing individuals who matched this description so we did the typical thing, found an overburdened local ASCE member. This individual was Bruce Rawls, our PNC representative and the County Utilities Director.

The chair should be an individual who can organize all the necessary details required to make a project such as this a success. The ability to regularly meet, track progress, and maintain a focus is essential. It is necessary to build a strong team of committed individuals (of course after the process is all over we all felt we should be committed for having taken on this project!).

Use your resources. Younger members, retired members, member spouses, members who have contacts or are employed by government agencies that are tied to the project are all necessary.

Many projects considered for recognition as a National Historic Civil Engineering Landmark can be large public works projects and will involve working with government agencies. The Inland Empire Section had much needed support from The Bureau of Reclamation, Department of the Interior, and the Corp of Engineers. The project would not have been successful had it not been for the help given by these agencies, their public relations representatives, and ASCE members within these organizations who helped gather the necessary information.

Establish contacts with those individuals who need to be involved. Our section made contacts with federal officials associated with the Dam, elected officials, and individuals still living who were associated with the project. This will also include ASCE staff, members and officers to arrange calendars and to get input. Remember the mantra, Start Early, Start Early, ….

Point number four, ***FOCUS ON THE REQUIREMENTS AND TIME CONSTRAINTS NECESSARY TO MAKE FORMAL APPLICATION THROUGH ASCE***. The Inland Empire Section noted that this activity was more complex than it first seemed; we likened it to putting together an EIS.

Begin immediately to gather the application forms and focus on what you need to make application. For Grand Coulee, one of our biggest challenges was spending hours researching this truly awe inspiring piece of civil engineering workmanship. Many of the needed facts for the application could be retrieved in a relatively short period of time but, hours later, we found ourselves still in the library reading about the history of the project and the individuals associated with it!

Point number five, ***PURSUE THE PROJECT WITH THE MINDSET OF CREATING A SHOWCASE FOR ASCE AND CIVIL ENGINEERING IN GENERAL***. Look for individuals to involve in the ceremony. This includes local elected officials, state and national representatives, those who may have participated in the project and those who have recollections of the project. Be sure to include all the media: television, radio, newspapers. We found it beneficial to prepare a press kit that specifically explained the time, location, and details of the dedication.

The Inland Empire Section was fortunate to have a past section president who worked on the project and still lived in the area. Mr. L. Vaughn Downs was a young project engineer on Grand Coulee and had actually written a book about his experiences. Even though in his nineties, Mr. Downs spoke with much strength and passion about the Grand Coulee project both at a dinner the night before the dedication and at the dedication. Local television stations covered the dedication and interviewed Mr. Downs. The coverage was very positive and a good representation of Civil Engineering.

One of the most successful ways we had to create a positive view of ASCE was the dedication program. Don't overlook this simple item as a great tool to convey the message of ASCE. The program we prepared included interesting and historical pictures of the Dam and a short history of ASCE and our local section. Many members commented on how nice it was to have the short history of the local section.

Point number six, ***SELECT A PROJECT HISTORIAN/SECRETARY***. While we have been talking about past history and recognizing past projects and accomplishments, we sometimes forget we are creating history now. We are often frustrated as we research old projects. The general information is almost always there, but the details, the things that make a project come to life, are often missing or forgotten. By selecting a good historian/secretary, you can capture the details of your efforts and preserve the history of your actions. This will provide the details, the "life" of our actions, for the next generation of ASCE historians.

I'm sure that the engineers who worked day in and day out in severe conditions on Grand Coulee often thought their efforts were mundane and boring; yet today we scour libraries to find any available details about the day to day activities of this project because we find it so fascinating.

So remember: take good notes and record anecdotes because someone from ASCE will ask you to report on it and it is better to have a record of information months after the fact.

85th Anniversary of the Seattle Section

By J.E. Colcord, L.M. ASCE[1]

Abstract

The paper reviews the work of the Seattle Section since formation in 1913, until the present. The concerns of the Section are education, scholarship, professional practice, history, and committees--participation and structure.

Introduction

On 30 June, 1913, Samuel H. Hedges, Ernest B. Hussey and Joseph Jacobs organized the Seattle Association of ASCE. Samuel Hedges became the first president, Ernest Hussey the vice-president, and Joseph Jacobs the secretary.

Prior to that, the American Society of Civil Engineers (ASCE) recognized the growing importance of the Pacific Northwest and held its annual convention in Seattle at the Hotel New Washington from June 21-24, 1912. That convention marked only the second time that the Society had a convention in a western state. Several hundred delegates arrived by train, some via a special open flat car ride through the Rockies. The delegates were welcomed by Mayor Cotterill and treated to a "Alaskan lantern slide presentation," an address by William Mulholland on the Los Angeles Viaduct [sic], and an illustrated address by Asahel Curtis on Mt. Rainier. They also took "automobile rides" and sight-seeing trips that demonstrated the magnificent water power resources of the region.

----

[1]Presented at the Boston Meeting, October, 1998.
Professor Emeritus, Dept. of Civil Engineering, University of Washington, Seattle; Past president, Seattle Section; 132 More Hall, Box 352700, Seattle, WA 98195-2700, Tel. (206) 525-6162.

Locally, on the education front, the University of Washington was founded in 1861, as the oldest state-assisted institution of higher education on the Pacific coast. Civil Engineering courses were first given in 1874 by Professor Frederick Harrison Whitworth and later by Joseph Taylor (Math and Astronomy) and John Hayden (Military Science and Math). These courses were given in the Territorial University Building at the original downtown site. It was said that "these courses were popular with the students because the faculty knew what they were talking about." At Washington, Civil Engineering officially became an academic discipline in 1898 with the appointment of Almon H. Fuller as a full-time Professor of Civil Engineering. He had degrees from Lafayette and Cornell. The first faculty to become president of the Seattle Section was William Allison who taught at Washington from 1913-1917 and became president in 1923. By 1920 the University had graduated 136 students with a B.S.C.E. degree, including one woman, Adelaide Cooper ('06). Of those students, 4 became president of the Seattle Section. Today 19 graduates and 11 faculty (including 6 graduates) have served as Section president--one of the ties that lessens the perceived "town & gown" problem.

Education

The University of Washington Student Chapter was organized in 1921, after the ASCE National Board of Directors authorized chapter formation in the same year at schools with a four-year accredited program in civil engineering. Later in 1964 a Student Club was formed at Seattle University, becoming a Student Chapter in 1986.

The Seattle Section has made great efforts to assure the student chapters of professional assistance and guidance in areas such as curricula, programs, award evaluations and financial help. Chapter members officiate at events such as the Concrete Canoe (the UW won the regional competition in April 1998 with an outstanding entry named "Centennial"). The Daniel W. Mead Prize and other student paper competitions are well supported by Seattle Section members. Incidentally, a U of W student won in 1982.

The Engineering Education Committee was formed in 1965 and has been relatively active on academic matters. It went through several name changes such as Curriculum Advisory and finally as Student Advisory Committee to better describe its function. This committee acted more as a fire-department--ready to serve if needed during such times as ABET accreditation for the two schools. The day-to-day work with student chapters has been accomplished through the Student Faculty Advisor and the Section Contact Member.

Students are invited to serve on Section Committees and receive half-price dinners at the Section meetings. One meeting each year has been presented by students with papers from inter-university competition and usually with a lively presentation on the Concrete Canoe--from design to disaster or even a win, while the judges deliberate on the paper merits and determine the annual Student Paper competition winner.

Scholarship

One of the most important efforts of the Section has been in the area of scholarship. The Richard Williamson Jones Memorial Award, given to a native-born U.S. citizen of junior standing living in the Seattle Section area, has been presented at an ASCE Section meeting. The money was made available through a grant by Kathie Jones in memory of her husband, Richard Jones, Executive Vice-president of Generation Construction who was killed in an air crash in 1959. Remembering his own student experiences, Richard felt that some funds often made the difference to a struggling student, and the Section is proud to help in this award.

The Section, thanks to the work of several members, set up the R.H. Thomson Memorial Award in 1991. This award is given to entering Civil Engineering students and may continue through their student career. It was named for R. H. Thomson, City Engineer 1882-1887, 1892-1911 and 1930-1931. He was responsible for holding the key role in achieving essential public works for logical development of infrastruture of a great city--including City Light, Cedar River Water Supply, Denny Regrade, the North Trunk Sewer, and the Chittenden Locks. At present, 6 students have received this award from the Section.

The Section contributed to the H.M. Chittenden Memorial Award, presented to a student interested in surveying. Hiram Chittenden, the son of General Chittenden, the engineer in charge of the construction of the locks joining Lake Washington and Puget Sound and bearing his name, was a faculty member at Washington from 1923-1965, a true gentleman and active in the Seattle Section.

Currently, the Section is in the process of establishing an award in honor of our Honorary Members (Colin Brown, Charles H. Norris, William Shannon, Stanley Wilson) of ASCE who were affiliated with the Seattle Section.

Clearly, the Seattle Section has done well in stressing this important area, and the continuance of excellence in education will enhance the profession in years to follow.

Professional Practice

Under this title are the contributions of the Section to the profession and the community. In 1982, the Section took the initiative to be involved in a new King County Ordinance dealing with the allocation of contracts for professionals from minority firms. The Section has continuously made recommendations on appointment of public officials to office and boards relevant to Engineering. Through the Legislative Committee, the Architects and Engineers Legislative Council, and Washington State Board of Registration the Section has kept abreast of legislation affecting engineers and engineering practice and offered position papers on these subjects. In 1943 the Section was heavily involved in a proposed Amendment to the Seattle Section constitution that created the possibility of a collective bargaining unit. In 1980 the Section helped convince the City of Seattle administration that the new high-level West Seattle Bridge should be designed by a highly qualified local firm. This was after some questionable practices, such as those associated with Vice President Agnew, had created a local scandal. Both the high-level and low-level bridges were done locally and achieved great success.

In the early days, members such as General Hiram M. Chittenden, head of the Corps of Engineers in Seattle, pushed for and engineered development of the Seattle harbor. This cooperation continued on the location of the Metro expansion of the West Point Secondary Treatment Plant and the location of I-5 through the city, along with the construction of the second Lake Washington and Hood Canal floating bridges.

Perhaps the most outstanding role of the Seattle Section is that of making available to the public the special technical skills of the members. This April, as an example of the "New Public Relations Program," a seminar, open to the public, was held on "Landslides in the Puget Sound Region."

Another popular event, while perhaps not as technical, is the Popsicle Stick Bridge Competition put on by the Associate Member Forum. This is a very popular event which highlights the Civil Engineering profession to young prospective engineers. The Third Annual event, held in February of this year, was a smashing success and had participation of 37 area high schools and six celebrity judges, including John Stanford, Superintendent of Schools. Special recognition went to Olympic High School of Bremerton for a bridge holding a record 1,272 pounds. All 37 participating teams were truly winners as they learned through hands-on experience.

History & Heritage (H & H)

In 1982 the Section worked to have the Snoqualmie Falls Cavity Generating Station declared a National Historic Civil Engineering landmark. The H & H Committee of the Section has had recent success by having the John Stevens', Great Northern's Stevens' Pass Railroad Tunnels and Switchback System (1892) designated in 1993, and the H.M. Chittenden Locks and Ship Canal, designated in 1997. The committee is currently working on several local, regional, and national projects, such as the Cedar Falls Hydro project--the first municipally-owned lighting project in the country.

Project of the Year

The Seattle Section was involved in the Outstanding Civil Engineering Achievement (OCEA) awards since 1977, when the Award of Merit was presented to the Seattle Freeway Park. In 1987 Seattle received the OCEA regional nomination award for the Metro Renton ETS Outfall by the Pacific Northwest Council. In 1981 the National Award of Merit was given to the Naval Submarine Base at Bangor, WA, located in the Kitsap Branch area. The National Outstanding Civil Engineering award was presented to the Seattle Section for the Mount Baker Ridge Tunnel/Lid Complex in 1990, and again in 1992 the Section was given the National OCEA award for the West Seattle Low Level Bridge with its unique hydraulic rotation scheme.

In addition, in 1984 the Seattle Section set up the Local Outstanding Civil Engineering Achievement (LOCEA) award. This "small" project award must be under the direction of a Seattle Section member, construction cost must not exceed $1.5 million, and the work must have been completed in the calendar year. These award recipients receive a bronze plaque, and the government agencies or private ventures involved have been very receptive of the Small Project award as it is more humanizing and personal than awards for major projects, and truly represents Civil Engineering to the public. In 1997, as an example, the LOCEA award winner was the Runway Improvements at the Snohomish County Airport, a project designed by Reid, Middleton Consultants of Edmonds, WA.

National ASCE Committee Service

The Section is in National Zone IV, District 12, and has been well represented on National Committees--for example in 1990-1991 at least 85 Section members served in this capacity. In the 1998 ASCE Official Register, Larry R. Wade, (F. ASCE) serves as Treasurer on the National Board of Directors. He was Director from 1996-1997. In the past, the Section has had other Directors such as

Eugene McMaster (1983-1984) and Professor Fred Rhodes (1959-1961). Some members affiliated at one time with the Seattle Section and became President of ASCE are James Poirot (1994) and John Stevens (1927).

Locally, the Section committees, which are the true life-blood of the Section consist of the following in 1997-1998:

A.M.F. (Associate Member Forum)
House and Hospitality
Geotechnical
Audit
Water Resources and Environmental
Student Advisor - Seattle University
University of Washington
Membership (chaired by the President Elect)
Puget Sound Engineering Council Representative (one of the Section Directors)
Legislative
History and Heritage
Transportation and Infrastructure
Management in Engineering
Washington State Board of Registration Rep.
Structural/SEAW Liaison
AELC Rep.
Waterways
R.H. Thomson Scholarship
Transpeed
Lifeline-Earthquake
Budget (Treasurer Chair)
Program
Convention 2000
Professional Practice

The Section is run by the elected officers. The officers consist of:

President
President-elect
Past President
Secretary
Treasurer
3 Directors with overlapping terms
AMF Representative
AMF President

We also have two Branches: Kitsap and North, that better serve our members who reside and work outside the Seattle-Bellevue metropolitan area. The *Monthly*

Newsletter is the organ keeping the 2300 members informed and contains the meeting notice, President's column (including each Branch), Board minutes, Community meeting calendar, and special relevant information for the month.

The Section periodically published a Membership Directory. This directory included the Section Constitution, lists of past officers, history, and ASCE benefits, telephone contact numbers, and Student Chapter information. Recently, it has become increasingly difficult to have an up-to-date hard-copy directory due to job and telephone area code changes. Thus, the latest information is now planned to be available on the web site:

www.asce.org/gsd/sections/seattle

Conclusion

As a general comment, it is interesting to compare a meeting of 49 years ago with the present. I recall interminable discussion at the old "Engineers Club" in downtown Seattle about local dues--a raise of 10 cents caused fierce discussions from the 20 or so members present. Now, the meetings are more technical and 100-150 members attend with the business conducted by the Executive Committee with only the occasional necessary membership vote. I also believe that, through the extensive and varied committee structure and the AMF, recent graduates have more say than they did at the time of the established "Old Boys' Network" that I experienced--of course "I are one" as they say. One of the best things is the active participation of women--they are some of the true leaders of the Section, as predicted from their work in the Student Chapter.

References

ASCE, *Civil Engineering*, July, 1992
ASCE, *Guide to History & Heritage Program,* 1988, Revised
ASCE, *Membership Directory* (1991-1993), Seattle Section (includes 75th Anniversary History)
ASCE, *Monthly Newsletter of the Seattle Section* (various)
ASCE, *Official Register,* 1996, 1998
Seattle Times, June 21, 1912

HISTORICAL CIVIL ENGINEERING PROJECTS IN TEXAS

Martha F. Juch, P.E., M.ASCE[1], Steven P. Ramsey, P.E., M.ASCE[2],
Jerry R. Rogers, Ph.D., P.E., M.ASCE[3], and Fred P. Wagner, Jr., P.E., M.ASCE[4]

The Texas Section of the American Society of Civil Engineers was founded in 1913 and celebrated its 85th Anniversary in 1998. As of January 1998, eight National and thirteen Texas Civil Engineering Historic Landmarks in our state had been recognized by the American Society of Civil Engineers. These are presented herein as reminders of the engineering achievements of the past and as inspiration for the engineering achievements yet to come.

The oldest known civil engineering project in Texas is El Camino Real, which means "The King's Highway" or, more literally, "The Royal Road". El Camino Real is a 16th century Spanish transportation artery which provided support, defense, and political stability to early colonists. Comprised of three branches, the El Camino Real originated in Mexico and passed through all of the southern territory of the present United States. The path of the Western Branch of El Camino Real extends from Durango, Mexico up through Arizona through San Diego to Sonoma,

---

[1] Manager, Advanced Planning Unit, Harris County Flood Control District, 9900 Northwest Freeway, Houston, TX 77092

[2] Chief Engineer, San Antonio River Authority, P.O. Box 830027, San Antonio, TX 78283-0027

[3] Department of Civil and Environmental Engineering, University of Houston, Houston, TX 77024

[4] 5230 18th Street, Lubbock, TX 79416

California. The Central Branch originates in Veracruz, Mexico and passes through Chihuahua and El Paso before ending in Santa Fe, New Mexico. The Central Branch of El Camino Real was named a Texas Historic Civil Engineering Landmark in 1988 by the Texas Section and is marked by a plaque in El Paso. The Eastern Branch of El Camino Real ran for 2400 miles between Veracruz, Mexico and Saint Augustine, Florida and was begun by Hernando Cortez in 1519. Many of today's streets and highways follow the original path of the Eastern Branch of El Camino Real through cities in Mexico and the southern United States as it passed through San Antonio, Nacogdoches, Natchez, Pensacola, and Tallahassee. The Eastern Branch of El Camino Real was named a National Historic Civil Engineering Landmark by ASCE in 1986 and is represented by a plaque in San Antonio.

The second oldest National Historic Civil Engineering Landmark in Texas is the Acequias of San Antonio. The acequias formed a rudimentary canal system for the San Antonio mission settlements and were the primary water supply for the first colonists in Texas. The first of eight was constructed in 1718, and portions of the Espada Acequia and San Juan Acequia are still in use. Built between 1740 and 1745, the Espada Aqueduct spans a small creek and is the only such structure in the United States. This aqueduct has provided irrigation water to neighboring fields for over two centuries. The Acequias of San Antonio were named a National Historic Civil Engineering Landmark by ASCE in 1968.

The third oldest National Historic Civil Engineering Landmark in Texas was actually the second to be recognized by ASCE. The International Boundary Marker No. 1 establishes the initial survey reference point marking the official international boundary between the United States and Mexico and was designated a National Historic Civil Engineering Landmark in 1976. Located on the west bank of the Rio Grande between El Paso and Juarez, Mexico, the monument marks the easternmost end of the land boundary (the Rio Grande forming the rest of the eastern border) and, hence, of the survey which extended westward to the Pacific Ocean. The marker stands as a monument to the professional skills of the Emory-Salazar Commission, the American surveying team that determined its location in 1855. Built the same year, the four-sided marker is situated with its south half in Mexico and its north half in the United States. The stone monument is 12 feet high and 5 feet wide at its base. In 1966, it was refaced with white marbleized concrete, and a concrete base was installed.

The fourth project in Texas to be named a National Civil Engineering Landmark by ASCE is the Houston Ship Channel. Designated by ASCE in 1987 and dedicated at the Fall Section Meeting in Houston, the plaque recognizing the Ship Channel as a National Landmark is located on a monument at the docking facility/visitor's center for the passenger vessel operated for public tours of the channel. The Houston Ship Channel contributes approximately $3 billion annually to the area's economy and is the center for industrial/ commercial/ transportation activities serving the City of Houston, the State of Texas and the United States. The Houston Ship Channel is a 50-mile-long waterway that stretches inland from Galveston Bay and the Gulf of Mexico to the Port of Houston. This facility changed a small town on the banks of Buffalo Bayou into a port ranked eighth largest in the world and second largest in the nation. Although some improvements to the Bayou began as early as 1839, it was not until 1909 that Harris County citizens elected to form a navigational district, the Port of Houston Authority, charged with supervising the port and issuing bonds to help fund channel dredging. On November 10, 1914, the port was finally opened to deep water navigation. Subsequent work has enlarged the channel to its present average depth of 40 feet and width of 400 feet.

The San Jacinto Monument in Houston celebrated its golden anniversary in 1989 and has become a symbol of the spirit of Texas which is recognized across the United States. Along with the San Jacinto Museum of History which forms its base, the 570-foot monument took three years to complete and was dedicated on April 21, 1939. The monument commemorates the Texas Army's victory in 1836 that gained independence from Mexico and eventually added a million square miles to the United States. Alfred C. Finn, a Houston architect, designed the monument and commissioned William McVey, a faculty member of Rice Institute (now Rice University), to sculpt the friezes surrounding the base. The engineering design was by Houstonian Robert Cummins. W. S. Bellows Construction Company, also of Houston, was in charge of construction. The plans called for an octagonal shaft, faced with Texas Cordova shellstone, to rise above the battleground from a 125,000-square-foot base and concrete mat foundation. A traveling form was used to place the shellstone as construction moved up the column. The monument is the central focus of the San Jacinto Battleground State Historic Park and was designated as a National Historic Civil Engineering Landmark in 1992.

Dedicated as a National Civil Engineering Historic Landmark in 1993, the Denison Dam is on the Red River (the Oklahoma-Texas border) near Denison Texas. Completed during World War II (1943), this dam was the largest rolled-earth fill dam in the world. The dam was authorized by an Act of Congress on June 28, 1938. Coffer dam closure on the Red River occurred on July 25, 1942, the project was fully operational for flood control in January, 1944, and the facility was dedicated in 1945. The rolled earth feature of this project was unique and its successful completion established an engineering approach that became the standard for construction of most subsequent earthfill dams. Testing equipment developed during the construction of this project also became the standard for other similar projects.

The seventh project in Texas to be named a National Civil Engineering Landmark by ASCE is San Antonio's River Walk and Flood Control System. This landmark demonstrates a pioneering approach to engineered solutions by establishing flood control while respecting the natural beauty of the San Antonio River. Studies in 1865 and 1921 recommended solutions for San Antonio River flood problems while projects that developed from the 1920's to 1940's achieved a number of the early recommendations. Significant achievements include the construction of Olmos Dam as a detention reservoir in 1927, and construction of the "Great Bend Cutoff" in 1929. In addition to flood control the River Walk was constructed between 1939 and 1941 as a project that offered a genuine respect to beauty and the historical and environmental resources of the community. Recognized by ASCE in 1996, San Antonio's River Walk and Flood Control System is truly a project of historical significance.

The most recent historic civil engineering project to be recognized by ASCE is the Texas Commerce Bank Building (locally known as the Gulf Building). The building is located in downtown Houston and was opened to the public on April 16, 1929. Construction on the building began in 1927, at which time engineer W. E. Simpson recommended the use of a mat foundation. Consolidation tests were made and Mr. Simpson's recommendations were confirmed by Charles Terzaghi. Steel construction was up to the 14th level when the owner requested an increase of four floors for a total height of 36 floors. The extensive use of welding to attach angles to the riveted H columns in order to carry the additional loading was the first use of welding for steel frame fabrication in Houston and probably the nation. Another unique feature of the building was the use of a split beam-tee section with driven rivets as connectors to serve as the moment connection, which reduced the floor-to-ceiling heights.

In addition to the eight Historic Civil Engineering Landmarks recognized by national ASCE, the Texas Section of ASCE has designated thirteen projects as Texas Historic Civil Engineering Landmarks. Table 1 lists these landmarks and shows the wide variety of historic projects in our state. One of these, the El Camino Real-Central Branch, was described above and serves to illustrate one of the many projects in Texas which, in the years and decades which followed its completion, not only affected the economy of the entire state, but also the economy of a larger region of our nation and our international commerce with Mexico.

The Waco Suspension Bridge not only is the second oldest local landmark to be recognized by ASCE in Texas, but also has the distinction of being the first project to be designated a Texas Historic Civil Engineering Landmark by the Texas Section in 1971. At the time it was constructed in 1870, the 475-foot Waco Suspension Bridge was the only bridge across the Brazos River and was the longest suspension bridge in the world. The steel trusses and wire cables were procured in New Jersey from the John Roebling engineering firm, which built the Brooklyn Bridge a few years later. The trusses and cables were brought to Galveston by steamer, relayed by rail to Bryan, and brought to Waco by oxen. Three million locally manufactured bricks were used in the structure. The bridge removed a major obstacle to the emigrants traveling westward and opened land to the west of the river to a vast migration movement. Waco soon became a major transportation and manufacturing center and was on one of the principal cattle trails of the era. By charging five cents per head "for each loose animal of the cattle kind", the bridge's original burden of debt was soon relieved (Roger N. Conger, "Waco: Cotton and Culture on the Brazos", Southwestern Quarterly, Vol. LXXV, No. 1, July 1971, pp.54-58).

The third oldest Texas Historic Civil Engineering Landmark is the Alamo Portland and Roman Cement Works in San Antonio. Completed in 1880, this plant is recognized as one of the pioneer Portland cement works in America. The Alamo Portland and Roman Cement Works was the first Portland cement factory in the American Southwest and has been acclaimed as the first Portland cement factory west of the Mississippi River in the United States. The factory was designated a Texas Historic Civil Engineering Landmark by the Section in 1976. Recently the factory was razed and replaced by a retail shopping center, although the factory's towers, which have graced the skyline for more than 115 years, were preserved.

Also designated a Texas Historic Civil Engineering Landmark in 1976, the Franklin Canal was constructed between 1889 and 1891 and stretches for 30 miles down the El Paso Valley from the City of El Paso, Texas. The canal was the first large-scale irrigation effort undertaken in extreme west Texas and, much like its precursor, the Acequias of San Antonio, it has provided a dependable supply of irrigation water to the region which it serves. Originally constructed by private enterprise, the canal was acquired by the U.S. Reclamation Service in 1912. The Franklin Canal has served to promote the economic growth of the region and has been a key factor in its development.

The supplying of water for cities is one of the fundamental areas of the civil engineering profession. The Holly Pump Station and North Holly Water Treatment Plant provide an early example of this important service. The Holly Pump Station was completed in 1892 and the treatment plant in 1911. Some of the original facilities, including the pump building and sections of the filter and chemical buildings, are still in use. The original 8 MGD steam pumps were some of the largest built at the time they were installed and they served the city of Fort Worth for nearly 60 years. The rapid sand filter treatment plant was the third such plant in Texas and the largest in the southwest at the time of construction. The treatment plant is still an integral part of Fort Worth's water system and was dedicated by the Texas Section as a Historic Landmark in 1992.

The sixth Texas Historic Civil Engineering Landmark listed in Table 1 is the Galveston Seawall and Grade Raising. Initial construction of the Galveston Seawall began in 1902 as a result of the devastating storm of 1900 in which over 6,000 persons were killed. The construction of the wall to a height of 17 feet above mean sea level, and the accompanying raising of the overall elevation of the City of Galveston from eight to fifteen feet, constituted the first such large scale seawall/grade raising effort on the Gulf Coast. The present seawall is 54,790 feet in length and was constructed by Galveston County and the U. S. Army Corps of Engineers. The Galveston Seawall and Grade Raising was designated a Texas Historic Civil Engineering Landmark by the Section in 1975.

The Mills Building in El Paso is the first commercial building to be designated a Texas Historic Civil Engineering Landmark. Named a Historic Landmark by the Texas Section in 1981, the building merits this distinction by being the first large reinforced concrete structure constructed in Texas and among the earliest skyscrapers of its type in the United States. The structure construction was started in 1909, initially completed to eight stories in 1911, and ultimately completed to twelve stories in 1915. It was designed by General Anson Mills (1834-1924), an early El Paso pioneer and a civil engineer. The Mills Building has long been a landmark depicting the earliest days of El Paso's history and stands in an area which has been revived to save the architecture of other similar structures.

Two unusual but similar structures in the Dallas-Fort Worth area have received the distinction of being named Texas Historic Civil Engineering Landmarks. The Houston Street Viaduct in Dallas and the Paddock Viaduct in Fort Worth were each completed in the early twentieth century and contributed greatly to the economic growth of he area. The Houston Street Viaduct is the oldest of the two, being completed in 1911 and spanning the Trinity River as a connection between the Dallas Central Business District and the early suburb of Oak Cliff. The 6,562-foot Houston Street Viaduct is one of the longest built with reinforced concrete arches. The Houston Street Viaduct was designated a Texas Historic Civil Engineering Landmark by the Texas Section of ASCE in 1989.

The Paddock Viaduct crosses the Trinity River in downtown Fort Worth. Construction began on the viaduct in 1912 and was completed in 1914. This structure was the first large concrete arch bridge erected in the United States using self-supporting reinforcement. The Paddock Viaduct served as the prototype for similar bridges across the United States which used this construction technique. The Texas Section named the Paddock Viaduct a Texas Historic Civil Engineering Landmark in 1976.

The Original Dallas Floodway is the second major civil engineering project in Texas to be conceived in the aftermath of a devastating natural disaster. Similar to the Galveston Seawall and the flood of 1900, the Original Dallas Floodway resulted from a 1910 plan for reclamation of the Trinity River Valley within Dallas following the record 1908 flood. The plan was revised in 1920 by Myers and Noyes Engineers. The construction of the project was completed by the City and County of Dallas Levee Improvement Districts in 1931 and reclaimed 10,553 acres with approximately 23.5 miles of levees and four pump stations. The Floodway was designated a Texas Historic Civil Engineering Landmark by the Section in 1989.

Buchanan Dam was dedicated as a recipient of Texas Historical Civil Engineering Landmark status in 1990. Originally the Hamilton Dam, the structure on the Colorado River was conceived in 1926. With plans prepared by Fargo Engineering Corporation in 1930, initial construction was begun by Fegles Construction Company in 1931. After several changes of ownership, the Lower Colorado River Authority (LCRA) was created by the Texas Legislature and assumed responsibility for the dam. The dam was renamed the Buchanan Dam after U.S. Representative J. P. Buchanan and was completed by LCRA in 1937. The dam is located in Burnet and Llano Counties, 12 miles west of the City of Burnet. It is the longest dam in the United States which uses a multiple arch concrete buttress section and is the only dam of this type in Texas. The dam is 10,871 feet long with a structural height of 145.5 feet. The reservoir formed by the dam is 30 miles long, 5 miles wide and has a surface area of 23,060 acres.

The Corpus Christi Seawall has the distinction of being the"youngest" Texas Historic Civil Engineering Landmark listed in Table 1. The 11,000 foot long Corpus Christi Seawall was conceived in 1921 and completed in 1941. It is the only storm tide protection project in Texas financed with state ad valorem tax remission funds. The seawall protects Corpus Christi from tropical storms, ensures the City control of its waterfront, and provides a strip of hydraulically filled property built in the 200 to 350 feet between the seawall and the former shoreline. The firm of Myers and Noyes Consulting Civil Engineers designed construction plans for the undertaking. The project was the first major reinforced concrete step-type seawall constructed with traveling metal frames. The Corpus Christi Seawall was dedicated as a Texas Historic Civil Engineering Landmark in 1988.

The Civil Engineering Historic Landmarks located throughout Texas serve as permanent documentation of the development of Civil Engineering in our state. Perhaps one of the most important of these is the monument to the founding of the Texas Section of ASCE located in Corpus Christi. Dedicated in 1983, the monument stands as a reminder of the dedication and resolve of the Texas civil engineers who first met in 1913 in order to form one of the first authorized Sections of the prestigious civil engineering professional society. Texas could not have developed into the strong and progressive state that it is without the continued support and vision of the civil engineering community. Recognition by ASCE of the historical significance of civil engineering projects in the state continually popularizes the profession and serves to increase public awareness of engineering in general.

TABLE 1
NATIONAL AND TEXAS HISTORIC CIVIL ENGINEERING LANDMARKS

Landmark	*Location*	*Year of Construction*
NATIONAL ASCE LANDMARKS:		
1. El Camino Real-Eastern Branch	San Antonio	approx. 1519
2. Acequias of San Antonio	San Antonio	1718-1745
3. International Boundary Marker No. 1	El Paso	1855
4. Houston Ship Channel	Houston	1909-1914
5. San Jacinto Monument	Houston	1936-1939
6. Denison Dam	Denison	1942-1943
7. River Walk and Flood Control System	San Antonio	1939-1941
8. Texas Commerce Bank Building	Houston	1927-1929
TEXAS ASCE LANDMARKS:		
1. El Camino Real-Central Branch	El Paso	approx. 1519
2. Waco Suspension Bridge	Waco	1870
3. Alamo Portland / Roman Cement Works	San Antonio	1880
4. Franklin Canal	El Paso	1889-1891
5. Holly Pump Station/N. Holly Water Treatment Plant	Ft. Worth	1892
6. Galveston Seawall and Grade Raising	Galveston	1902-1904
7. Mills Building	El Paso	1909-1915
8. Houston Street Viaduct	Dallas	1911
9. Paddock Viaduct	Fort Worth	1912-1914
10. Texas Section Founding Marker	Corpus Christi	1913
11. Original Dallas Floodway	Dallas	1920-1931
12. Buchanan Dam	Austin	1931-1937
13. Corpus Christi Seawall	Corpus Christi	1930s-1941

Highlights of A History of the Colorado Section of the American Society of Civil Engineers

Arthur J. Greengard, Jr., P.E., F. ASCE

The Colorado Association of Members of The American Society of Civil Engineers was organized in 1908. Engineers in Colorado have been meeting informally for the previous two years to plan for ASCE's 40th Annual Convention to be held in Denver. This was the first convention held west of the Mississippi River since San Francisco in 1896. Attendance was 440. The founding of The Association is attributed to Herbert S. Crocker, a consulting engineer, who had rose to National President in 1932.

The first meeting of The Association was January 9, 1909. The Association was the second local Association approved by The American Society of Civil Engineering predated by San Francisco in 1905. This is not to be confused with the Boston Society of Civil Engineering, organized in 1848 followed by the American Society of Civil Engineers in 1852, but not approved by the American Society of Civil Engineers since it did not exist. There were 53 members. Meetings were held the second Saturday of each month, except July and August. Dues were $2 per annum. Early meetings consisted of discussions of papers previously published in "Transactions of The American Society of Civil Engineers".

By May of 1909, discussions were held of papers presented by local members and involvement of the University of Colorado was secured. The National ASCE President was invited to the December 1909 meeting, for the first time, but could not attend. Engineering education was first discussed in January of 1910, with four Colorado engineering schools present; The University of Colorado, Colorado School of Mines, Colorado School of Agriculture and Mechanics and Colorado College. Legislative activities were also initiated in 1910 with respect to highway funding and water resources. Also, in 1910 membership recruitment was discussed. There were now 60 members and a budget of $200. Average meeting attendance was 17 members (maximum -24, minimum -8) and 12 guests (maximum -45, minimum -1). The Association prepared a Licensing of Engineers bill which was rejected by the legislature. In 1913, the Legislative Committee recommended that the State Engineer be paid $4,000 rather than $3,000 per year. The shortest term Association President was also in 1913, June 14 to 20th, due to absence from State. The Association received its first invitation to attend the National ASCE "Meeting of Presidents" in New York City in 1915. In 1914, Colorado was one of 13 State Associations. Also the United States Reclamation Service moved to Denver from Washington D.C. in 1915. The first joint meeting with The American Institute of Electrical Engineers was in May of 1916. During World War I, two active Colorado members were involved in construction of a army supply base in Brooklyn and powder plant in West Virginia.

In 1920, meetings were changed from Saturday to Monday. Crocker was secretary of National ASCE. The Colorado Association of Members became the Colorado Section. The Section adopted a resolution against metrication in 1922. Mr. A.O. Ridgway, a Section member, was nominated as District 11 Director in 1923, but was unsuccessful in his campaign. He was, however, bested by Mr. George Anderson, former Association President, now residing in California.

The financial status of the Colorado Section showed income of $160 in dues and $134 in allotments. Oil shale was discussed in 1925 at a cost of $25 per barrel to produce compared to $150 per barrel for oil. The annual national ASCE budget in 1925 was $210,000. In 1925, only 47 members of 125 paid the $2 Section dues and received the $2 allotment. A second national convention was held in Denver in 1927. Rooms at the Brown Palace were $3.50 to $6.00 per day. The student chapter at The University of Colorado had 60 members.

The 189th Section meeting was announced in January and February of 1930. Also, in 1930, a comment exists which states "Plan D is even worse in that it includes the State of Iowa with respect to redistricting". Colorado and Iowa are still included in District 16. R.C. Gowdy from Colorado was the first District 16 Director residing in Colorado. In 1933, ASCE issued an edict that in order to qualify for Section allotments, the Section must meet more than five times per year, must have technical meetings, and must send meeting notices. The November 1933 meeting was Reclamation Night signifying the influence of the Bureau of Reclamation on the Section at this time. Four reels of motion pictures were shown on the construction of Hoover Dam. There were 284 present. It was noted in 1935 that the Lincoln Tunnel, Fort Knox Gold Depository and Bonneville Dam were all WPA projects. Radio was also the subject of a meeting in 1935. The first of many Section annual reports, at least in existence, is 1934-1935. Six of the original 53 members were still active. The president stated "The engineer should be a leader and molder of public sentiment rather than a hirling". Not that this is new, there are many similar statements dating to 1908, but that is still stated with similar results in 1998. A history of the Section was written in 1936 which credits the founding of the Section "Over Highballs in a bar of the Denver Athletic Club after a game of golf". Mr. Roy Gowdy had ascended to Zone III Vice President. 1937 was also the first year of Life Memberships, with seven in

Colorado. In 1939, a Dr. Arthur S. Adam of the Colorado School of Mines stated "The engineer is not appreciated by the public as much as they should, largely because he is careless in his manner and dress." Also, a report from the Salary Committee said that the beginning salary of $170 per month as advocated in the February issue of Civil Engineering should be reduced to $125 per month.

The 70th Annual ASCE Convention was held in Denver in 1940, the third Denver Convention. ASCE now had 16,000 members. Ninety Section members participated in the convention. There were now 72 local Sections. Denver was called "American's Rock Garden". Commercial exhibits were frowned upon, but donations from commercial enterprises were not. Colorado had 571 registered professional engineers in 1940. There were 301 ASCE members, 157 paying local dues. The Bureau of Reclamation accounted for 73% of the junior members, 55% of the associate members and 41% of the members. Student chapters at CU and CSU were 65 and 28, respectively. By 1942, Colorado engineering firms were designing ordinance depots in Gallup, New Mexico and Flagstaff, Arizona and a bomb loading plant in Grand Island, Nebraska. ASCE proposed collective bargaining for engineers in 1944 - 1945, which the Section was violently opposed to. Planning for the 1952 centennial ASCE convention, hopefully to be held in Denver, began in 1947 with $200 "earmarked" for the convention. The 1947 Daniel Mead topic was "Is it ethical for a professional engineer to utilize the services of a nontechnically trained employee so that the latter is lead to believe that through training obtained in performing services, he may obtain a professional rating" with word limit 2000. Royce Tipton was endorsed for Zone III Vice President. Herbert Crocker died in 1949.

The Colorado Section suffered its first financial deficit in 1950 of $100. The 1952 ASCE convention was authorized in 1950 and local dues were raised from $2 to $3. The Section now had 531 members. Unfortunately, there is no record of the convention except "good attendance by local members is a primary factor in the creation of the friendly western attitude that is essential to the success of this convention". Walter Jessup, editor of Civil Engineering made a presentation in April of 1953 and Daniel Terrell, ASCE President in April of 1954. Denver Mayor Will Nicholson and Colorado Governor Ed Johnson were in attendance in September, 1955. Dr. Erkel of the University of Colorado was elected District 16 Director in 1955. The October meeting in 1956 featured Dr. Wilber Wright, grandson of the Wilber Wright. ASCE President Mason Lookwood also visited the Section in May 1957. The first meeting of the Grand Junction branch occurred in 1959 with 16 present.

William H. Wisley presented "What ASCE membership means to the individual" in February of 1960. A engineers building was also constructed in Denver in 1960, partially sponsored by the Colorado Section. This was the same time as construction of the United Engineering Center in New York City to which 185 Colorado engineers contributed $7,265. The Engineers Club featured bingo every second and fourth Friday, bridge every third Friday and

a smoker on September 29th. Colorado Section dues were raised to $4 in 1961 due to a lack of dues paying members. In 1962, a speaker announced "engineers spend too much time speaking to glorifying each other and not trying to convey their position to the layman". Royce Tipton was elected Zone III Vice President in 1964. The Engineers Club had a membership of 400 by 1966. There was also a national convention in Denver in 1966, but no record. June of 1967 was the 550th meeting of the Section. Also, in 1967, there was an accounting of Section members who became national officers: President Crocker (1932), V. P.: Crocker (1919-1920), Ketchum (1925-1926), Gowdy (1937 - 1938) and Tipton (1964 - 1965). Directors: Crocker (1915 - 1917), Ketchum (1918 - 1920), Anderson (1921 - 1923), Crawford (1935 - 1937), Tipton (1944 - 1946) and Eckel (1955 - 1958). Honorary members from Colorado were Crocker, Ketchum, Savage, Huntington, Eckel and Lane.

The Gunnison Tunnel and Uncompagre project became ASCE historical landmarks in 1971. Cheesman Dam became an ASCE historical landmark in 1973. The Colorado Springs branch was reactivated in 1974 with 57 members. Section dues were raised in 1974 to $5. Denver Mayor Bill McNichols addressed the Section in January, 1975. Another national convention was held in Denver in 1975. The 75,000 ASCE member was Shiril Voight, a University of Colorado graduate and structural engineer at Coors, from Pittsburgh. The 650th meeting occurred in February 1978. The Mini Brook Bill passed the Colorado Legislature in the summer of 1979.

1980 was the first time ASCE members could get a 25% discount on Hertz Rental Cars. Colorado Section dues rose from $5 to $8, with 1800 assigned Section members. Colorado also had 131 life members in 1980. In 1981 there was an all engineers dinner during engineers week with 425 attending. Jack Swigert, former Apollo 13 Astronaut was the speaker. Ken Hansen became a District Director in 1981. In 1982, Districts 16 and 17 were combined and became District 16. Denver was considered a serious candidate for ASCE headquarters relocation in 1982 along with St. Louis and Dallas, ASCE opted to stay in New York. There was an ASCE spring convention in Denver in May of 1985 with 1400 attendees. The last of long standing Colorado Section holiday dinner dances was in 1987. In 1988 there was a listing of 191 Section members active on 360 national committees. It was reported that ASCE has 106,504 members, 550 committees and 6500 committee members. The last sections dues increase occurred in 1989 from $8 to $10.

In 1990, ASCE published 25,000 coloring books. James Poirot was elected Zone III Vice President in 1991. He was elected President Elect in 1992, by petition. Colorado State University hosted the concrete canoe races in 1992. In 1994, the Section President lamented. "The average person does not understand our profession and has no idea of the enormous contributions civil engineers have made to the well being of society". Anything after 1994 is not to be considered history.

In conclusion, the Section is good at having meetings. There are also reoccurring themes throughout history. Social involvement comes to mind, all talk and no action. That is who we are. Why does it upset us so much? It appears that we all ought to be proud that we are not politicians. In review of this short paper, I find that these are not really highlights of the section. There are more interesting things in the book. But they are not low lights either. There are less interesting things in the book. Maybe they are just lights.

Robert E. Lee and the Saving of the St. Louis Riverfront

Jeff Keating, P.E., Member[1]

Abstract

In the 1830's St. Louis was the economic center of the American West, primarily due to its strategic position on the Mississippi River near the confluence of the Mississippi and the Missouri. The Mississippi River was meandering to the east, away from St. Louis' wharves and warehouses, threatening the city with economic ruin and possibly cutting trade routes to the west. Lt. Robert E. Lee of the Army Corps of Engineers was assigned the task of redirecting the river to preserve the harbor and the city using new techniques of hydraulic engineering.

In 1764 when Auguste Chouteau and Pierre Laclede arrived in the area that is now St. Louis, the site presented itself as an ideal location for a settlement. Unlike the nearby Mississippi River towns, the location was near enough the Missouri River to serve the fur trade, but was not in the floodplain. The west bank was the first high ground below the confluence of the Mississippi and the Missouri and had a gently sloped natural levee up to a bluff. This bluff left only those buildings located at the water's edge susceptible to flooding, an ideal arrangement for river commerce.

The region had been controlled at various times by England, Spain, and France. Even though the U.S. gained the area in the Louisiana Purchase in 1803, the young country did not exert full authority over the region until 1816. Illinois gained statehood in 1818, and Missouri followed in 1821. By that time, St. Louis had long been a strategically located city vital to the western frontier. The City of St. Louis was the base of navigation for the Upper Mississippi and its tributaries, and the head of navigation for the larger boats from the Ohio and Lower Mississippi. All the commerce of the Mississippi, Missouri, and Illinois rivers was concentrated in St. Louis, along with a large portion of the commerce of the Ohio River and the

[1] Bi-State Development Agency, 700 South Ewing Avenue, St. Louis, MO 63103, 314-982-1411, jkeating@BSDA.org

overland trails to the west. These trade routes brought to St. Louis' wharf "the products of every clime and every species of industry." (Drumm) The young city's importance was more closely tied to river commerce and the development of steamboating than to its size. Between 1817 and 1844 St. Louis' population grew from 2,500 to over 30,000, a 12-fold increase. During that same time river shipping increased from 2000 tons to 144,150 tons, a 72-fold increase. No city in the Union of similar size was so extensively engaged in trade.

Auguste Chouteau's 1780 map of St. Louis shows a relatively straight riverfront along the Mississippi's western shore. As early as 1823 it was observed that the river was meandering away from St. Louis and had formed two islands. (See Figure 1.) Bloody Island, named for the many duels fought there, was directly across from the city near the Illinois shore. The other troubling addition to the river was Duncan's Island near the city's south end. To make matters worse, a silt bar also formed between the southerly end of Bloody Island and the northern end of Duncan's Island, blocking access by riverboats during low water. These islands were not small and they were growing larger as time passed. In the 1830's the islands were each approximately a mile long (1.6 kilometers) and covered with cottonwood trees. Bloody Island was approximately 500 yards (500 meters) wide and Duncan's Island was about twice that. Both were growing toward the Missouri shore and Duncan's Island threatened to block access to the federal government's arsenal just south of St. Louis. If left unchecked the river would shift its channel east of Bloody Island and leave St. Louis, the economic center of the West, a landlocked city. In an age where river traffic was the primary means of commerce west of the Appalachians, this could mean the death of St. Louis. As riverboats had more trouble accessing the harbor, the fears of St. Louis residents mounted. The situation became so precarious that people refused to invest in real estate in St. Louis.

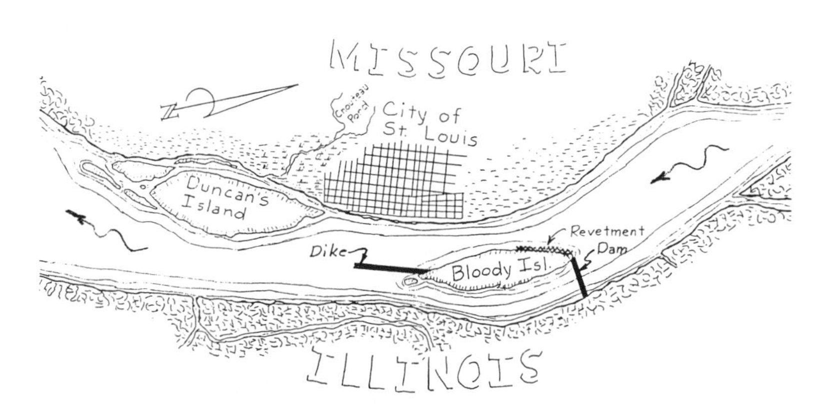

Figure 1.
Lee's Proposal for the St. Louis Harbor, 1837

While the city recognized the danger, it lacked the resources necessary to mount a viable effort to redirect the river. In 1833 the city hired John Goodfellow to plow the sandbars so that high water could more easily wash them away and to haul the sand away from the islands in an ox cart. Apparently, he dumped the sand into the river where it redeposited itself downstream, further blocking navigation. The river soon replaced the sand he removed leading to no reduction in the sandbars. The city spent $3000 of its $12,000 allotment on this effort before realizing that a more sophisticated approach was needed. As time went on, more riverboats were grounding in their attempts to land along more than half of the city's wharf.

During this time the debate raged over federal spending for public improvements. President Andrew Jackson, a strict constructionist, insisted that the Constitution did not permit federal funding for "internal improvements", arguing that this type of spending was the province of the states. Since St. Louis was an inland port, harbor improvements were considered a local problem, ignoring the regional and national importance of St. Louis' river trade and its effect on westward expansion. Luckily, Jackson was inconsistent in his application of this principle and spending on such improvements almost doubled during his tenure. In December 1833 the Mayor of St. Louis wrote to the House Committee on Roads and Canals asking for help in removing this hazard to the economic well being of St. Louis. The committee responded in its report that "a city so interesting should not be suffered to dwindle and decay if the interposition of legislative agency can prevent it."(Dobney) After a long battle, St. Louis was designated as a Port of Entry in 1836 and federal money was made available for port improvements. In that year Congress appropriated $15,000 to direct the river back to its former channel, but also stated that the United States Engineering Department had no officer available. Authority over the work was offered to the City of St. Louis. The city initially declined the offer, but later agreed to undertake the work "for and on behalf of the United States Government under the authority, direction and supervision of the Engineering Department of the United States."(Drumm)

Brigadier General Charles Gratiot, Chief of the Corps of Engineers and a native St. Louisan, personally examined the harbor in 1836 and stated that the problem could be overcome by constructing a wing dam from the Illinois shore to the head of Bloody Island and another from the foot of Bloody Island parallel to the Missouri shore. This would force the current west of Bloody Island and into Duncan's Island. Captain Henry Shreve was a brilliant engineer and the ranking Corps officer stationed on the western rivers. He was the inventor of the snag boat and developed the "cutoff" method of straightening S-curves in the river.(Shreve) His responsibilities included clearing the western rivers of snags and other obstructions to navigation. His territory included the Upper Mississippi, Missouri, Red, and Arkansas rivers. Gratiot discussed his plan with Shreve, who agreed the project was feasible. Gratiot placed the already overextended Shreve in charge of the work and Shreve, upon inspecting the harbor, stated that it was too late in the year to do any work until the next spring. He estimated $50,000 as the minimum sum that would be needed. An appropriation for an additional $35,000 was then approved by Congress.

Gratiot realized that the St. Louis harbor project was more than Shreve could effectively manage and began looking for another engineer to take the project over. Robert E. Lee was then a First Lieutenant working in the Washington office of the Corps of Engineers. He had graduated second in his class from West Point seven years earlier and was growing restless working in the Corps' Washington office. He had worked on such projects as surveying the boundary between Ohio and the territory of Michigan, averting what many feared might become an armed conflict.(Cooke) Lee was ready for his first assignment as a responsible supervising engineer. He volunteered for the St. Louis project and Gratiot assigned him to it.

General Gratiot sent a letter to J.B. Sarpy of St. Louis requesting that he act as Lee's business agent for the river projects and asking Sarpy to introduce Lee to the local dignitaries. This letter states that Lee had been assigned as the engineer for improvements to the Mississippi above the confluence with the Ohio. Proposed projects included improvements to the St. Louis harbor and devising a way to make the Des Moines rapids passable for river navigation.(Gratiot) Lee later wrote that he had also been directed to "do something to the Missouri, but they have given as yet no money and of course nothing has been done."(Lee, June 1838) Upon the arrival of Lee and his assistant, Second Lt. Mongomery Meigs, in St. Louis in August 1837, the river was still eight to ten feet above low water, but the rapids were reported to be at their lowest. They set off and, after gathering the necessary information to design a path through the rapids, returned to St. Louis by October 11. Lee then began his survey of the harbor. He filed a detailed proposal for improvements to the harbor with General Gratiot on December 6, 1837. Lee had studied the dike system used in Holland and was familiar with systems in place on the Hudson and Ohio rivers with design elements similar to what he proposed for St. Louis.

Lee's report to Gratiot included a detailed description of a harbor that was becoming too shallow for navigation and Lee's plan to correct the situation.

> *"The appropriation for the improvement of this harbor has for its object the removal of a large sand bar, occupying below the city the former position of the main channel of the Mississippi; which, gradually augmenting for many years, has now become an island of more than two hundred acres* (80 hectares) *in extent, covered with a growth of young cotton-wood, and reaching the lower part of St. Louis, which it shuts out from the river, to two miles* (3.2 kilometers) *below. The extensive shoals formed around its base extend on the east to the middle of the river, and, connected with the main land on the west, affords, at low water, a dry communication between. A flat bar projects from the upper end to the foot of Bloody Island, opposite the town, which, at low stage of the river,*

presents an obstacle to the approach of the city, and gives reason to apprehend that, at some future day, this passage may be closed."(Drumm)

Lee goes on to explain that the Mississippi is cutting a new channel east of Bloody Island and that the way to correct this is by restricting the channel against the Missouri shore. He proposed to do this by constructing a dam from the head of Bloody Island to the Illinois shore at right angles to the river, creating dead water that would redirect the channel. (Figure 1) By adding another dike downstream from Bloody Island, the water directed against Duncan's Island should wash it away. It would also be necessary to protect the head of Bloody Island to prevent the newly concentrated channel from eating-away its western shore. His estimate for this work on and around Bloody Island was $158,554, which included $63,574 for the dam at the head of the island, $80,680 for the dike below the island, and $14,300 to armor the west bank of the island. This was over three times the original estimate provided by Shreve, but apparently Lee's superiors approved of his plan and in June 1838 he began work with $50,000 in federal appropriations, an amount Lee said, "is about enough to commence (removing Duncan's Island)"(Lee, June 1838) and $15,000 approved in October 1837 by the City of St. Louis.

Lee and Meigs returned home for the winter and Meigs was reassigned to other projects. A young engineer, Henry Kayser, had worked with Lee and Meigs the previous year as a city employee. He was barely 21 years old when Lee arrived in 1837, but he now became Lee's key assistant on the harbor improvements. During the winter of 1837-38 he watched over the project in Lee's absence. He also took soundings of the harbor; an initiative that Lee roundly praised in a February 1, 1838 letter.(Lee, February 1838) During this time an ice jam developed at the shoal near the head of Bloody Island, forming a barrier that diverted water to the east channel, deepening it and diverting more flow to the Illinois side. Kayser reported to Lee in late February that the river channel was moving to the east. Lee wrote back from Arlington, "The direction which the channel has taken and its course East of Bloody Isd. has verified my fears on that Subject. I regret it very much as it will now be more difficult to change the current than before."(Lee, March 1838) On March 25, 1838 Robert E. Lee arrived in St. Louis with his wife and three children, having arranged to rent lodging in the home of General William Clark, the former co-Captain of the Lewis and Clark Expedition.(Wolferman)

Since Congress had not appropriated enough money to construct all the improvements Lee had proposed, he had to decide which of the improvements would provide the most immediate relief to the harbor. Lee determined that the dike at the foot of Bloody Island constructed parallel to the Missouri shore would have the greatest effect on Duncan's Island. Lee's design for this dike (See Figure 2), identical to the dam to be constructed from the head of the island to the Illinois shore, was largely borrowed from a design used on the Hudson River. It was to be ten feet (3 meters) wide on top, five feet (1.5 meters) above low water, and have side slopes of three horizontal to one vertical. Two rows of piles forty feet (12 meters) apart and five feet (1.5 meters) on centers were to be driven firmly into the river bed along the lines where the side slopes intersected the low water line. Brush was then to be interwoven to form a wedge-shaped mat, the mat laid at right angles with the dam and sunk to the bottom of the river with large stones. The area between the rows of piling was to be filled with smaller stones, sand, willow facines, etc. and slopes and crest covered with larger stones stood on edge and tightly fitted together to form a revetment from one and a half to two feet (0.4 to 0.6 meters) thick. The slopes below the low water line were to be secured by "throwing stones promiscuously over"(Drumm) and distributing them as equally as possible. Voids in the mats and facines would be filled by silt and sand trapped from the flowing water.

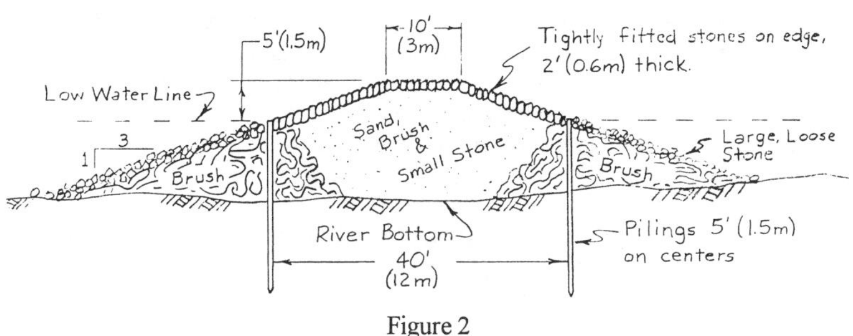

Figure 2
Cross Section of Dike and Dam

Construction work on the harbor improvements began in June and Lee was promoted to Captain on August 8, 1838. Capt. Lee personally directed the construction of approximately 2500 feet (760 meters) of dike from the foot of Bloody Island downstream. By October the harbor was beginning to look noticeably different. Duncan's Island was shrinking and the main channel had begun to deepen. Lee made it clear to city officials that, despite these short-term improvements, the dam from Bloody Island to the Illinois shore must be built. Without this dam a single flood or ice jam could permanently change the river's course. At this same time Lee became concerned that his original design would be susceptible to damage from ice flows, much as Kayser had reported the previous winter. Lee revised his plans and proposed the dam be made longer and intersect the Illinois shore farther upstream so it would present a slanting face to ice flows. (See Figure 3) He also became convinced that the longer dam might not be more expensive, if properly managed, since it would pass through a shallower channel.(Freeman) Congress adjourned on July 9, 1839 without acting on the request for further funding on the harbor. In order the get the most benefit from the work being performed, local funding was found, even though it made the young engineer an object of debate between local political factions and the press. When the political bantering was over, the City of St. Louis advanced $15,000 to pursue the work while conditions were favorable. General Gratiot approved Lee's use of this money and construction of the dam began. Lee began at the Illinois shore with the angled dike and drove a double row of piling 1300 feet (400 meters) toward Bloody Island. By early November the amount of running ice on the river forced work to be suspended for the winter even though not all spaces between the rows of piling had been filled with stone.

During that winter General Gratiot was abruptly dismissed from the Army due to financial improprieties. He had been a strong supporter of the St. Louis harbor and his successor, Colonel Joseph Totten, knew little of the project. Martin Van Buren, a stauncher strict constructionist, had replaced President Jackson in 1837, and the economy was in a downturn. With these changes, hopes of more federal money for the harbor were dashed. With the small amount of money remaining from the previous year's appropriation Lee set out in August 1839 to construct the dike to the head of Bloody Island. It was reported that Lee worked side-by-side with his men, sharing in the manual labor and common rations of the workers. This went to no avail because, before the year's work could be completed, an Illinois landowner secured an injunction against further work claiming diversion of the river would lower his property values. Lee was soon reassigned to use his experience to repair the defenses of New York Harbor, but not before he saw 700 feet (210 meters) of Duncan's Island disappear, the main channel deepen seven feet (2 meters), and riverboats accessing virtually all reaches of the harbor. Confidence in the permanence of St. Louis was restored and a building boom began.

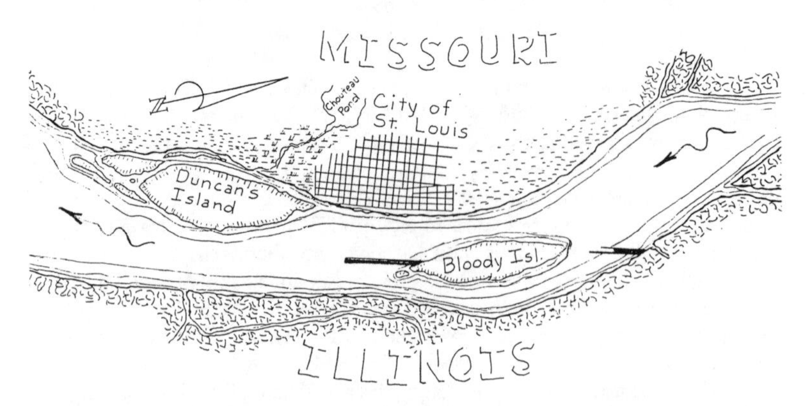

Figure 3
Condition of the Project When Lee Departed

Lee had expended only $57,000 of his original $158,000 estimate and the improvements were far from complete or secure. The upper dike was constructed only part way from the Illinois shore to Bloody Island – it needed to be completed and reinforced or the river could redirect itself. The piers were being undermined where they crossed shoals, being saved only by the stones lying at their base – they needed to be rebuilt and reinforced. There were several unanticipated scourings and depositions by the river that needed to be addressed before work already completed could be considered permanent. Lee reluctantly left St. Louis after turning the half-finished project over to Kayser to deal with it the best he could. Lee returned to St. Louis in 1840 to inspect the improvements and write a final report, but his direct responsibility for the project was over. His personal interest and involvement was not.

Henry Kayser had been working as Lee's assistant for the past two years and, knowing that Lee's time on this project was limited, had studied all Lee could teach him about river hydraulics and the harbor improvements. This approach now proved crucial to the ultimate success of the project. Kayser was named City Engineer shortly after Lee's departure and took the leadership role in construction of the dike at the head of Bloody Island upon the lifting of the injunction on October 7, 1839. Work continued for barely three weeks when weather forced work to be stopped for the winter. Kayser reported that the dike system, as then completed, was having the desired effect but it should be completed as soon as practicable, including a 600 foot (180 meter) cross-dike to strengthen the entire works. He was afraid, as Lee had been, that an ice flow or other such occurrence could destroy the improvements already in place.(Kayser) Lee maintained a keen interest in the St. Louis project and maintained an ongoing correspondence concerning the harbor with Kayser, who remained St. Louis City Engineer from 1839 until 1850 (except for a brief exception in 1846). Lee guided his former assistant in completing, repairing, improving, and expanding the original works in order to achieve the desired results. Lee and Kayser were both aware that hydraulic engineering was not an exact science and that adjustments to the design would be necessary once it was seen how the river behaved around the various dams and dikes. Some of this was seen almost immediately upon Kayser taking the project over. In June 1840 Lee wrote to Kayser concerning the undermining of part of the lower dike:

"My Dear Mr. Kayser

I have rec^{d} your letter of 31 May & Monthly Statements of funds rec^{d} &c. As regards the first, it gave me a very good idea of matters as they appear to be, but still not enough in detail to form a satisfactory opinion of the causes of the change of the current of the river. It seems to be very plain that the lower dyke is in a bad way, & probably its only safety will be in another deflection of the river during the June rise. Please let me know as soon as anything can be ascertained how it stands, and when the river falls sufficiently, how the Bars, shoals, Isds.; x channels &c are affected. Be as minute in your description as you can, & do not be afraid of either being misunderstood or tiresome. I am glad to hear that Duncans Isd. has been worn away some. Could you give me an idea of the quantity?"(Lee, June 1840)

The ongoing nature of the repairs and improvements to the dikes is demonstrated in a letter Lee wrote to Kayser in 1844, after one of the worst floods in recorded history. Lee wrote,

> *"I am very sorry to hear that the appearance of the condition of the Harbor is so bad. From your account I think that the river must have (cut) for itself another channel between the foot of the dike commenced at the Illinois shore the head of Bloody Isd. & that a larger quantity of water now passes East of B. Isd. than there has since 1839. You recollect the dike in question at that time extended across the small channel there existing & tended to divert a portion of the water west of B. Isd. I do not think that there will be any security for the Harbor until the pass East of B. Isd. is closed, & the water confined to the Missouri Shore at its lowest stage at least as far as the City extends. Whatever they adopt to effect this will in my opinion succeed. What is the best place they will have to find out. I hope they will discover it at last though it may take them some time."* (Lee, January 1844)

In all, Lee and Kayser kept up correspondence for over ten years after Lee left St. Louis. The efforts to keep the harbor open had an immediate and positive effect on St. Louis' economy. Steamboat arrivals doubled between 1840 and 1860 and the population had grown to almost 200,000. There has been some debate whether Lee designed the harbor improvements, or if Gratiot and Shreve, and later Kayser, should claim the credit. Superiors gave him direction and advice, but it was Lee's efforts, as described by Mayor John Darby, that led the project to success under his guidance. Lee was, "...with the hands every morning about sunrise and...working on his drawings, plans, and estimates every night till eleven o'clock.". After he left, it was his minute instructions to Henry Kayser that made possible the completion of the project over the next 16 years. Mayor Darby, who was also well acquainted with Gratiot, Shreve and Kayser, left no doubt that he credited Lee with the success of the project. "By his rich gift of genius and scientific knowledge, Lieut. Lee brought the Father of Waters under Control. ... I made known to Robert E. Lee, in appropriate terms, the great obligations the authorities and citizens generally were under to him, for his skill and labor in preserving the harbor. ... The labors of Robert E. Lee can speak for themselves."(Darby)

Lee's time in St. Louis also left a strong legacy of creative engineering. Henry Kayser became an engineer of note, primarily because of the training he received from Lee. There was also a young man who spent much of his time watching the progress of the work. James Buchanan Eads was an inquisitive 17 year-old when work began on the St. Louis harbor. He carefully observed Lee's approach in redirecting the river by spending much of his spare time at the riverfront observing the work when not busy with his job as a clerk in a dry goods store. He later applied these lessons to a series of monumental problems that preserved his place as one of the most inventive people in U.S. history.(Barry) These accomplishments included designing the first successful diving bell, building a fleet of ironclad river ships that helped the Union secure the Mississippi during the Civil War, and building his famous bridge over the Mississippi at St. Louis. He was concerned that the scouring action Lee had used to his advantage in redirecting the river would undermine his bridge if the piers did not go to bedrock. But nowhere did Eads follow in Lee's footsteps so closely as when he designed the South Pass jetties to carve a navigation channel though the alluvium clogged mouth of the Mississippi River. These jetties were quite similar in design and construction to the dikes and dams Lee used in St. Louis.(Barry)

Work on the St. Louis harbor continued over the years until its general completion in 1856 at a cost of $175,000. Duncan's Island was washed away entirely and Bloody Island rejoined the east shore and is now part of East St. Louis, Illinois. At extreme low water, pilings are still visible in the vicinity where the dams and dikes were constructed. The river is where early American engineers decided it would be; and it is sound engineering, hard work, and courage that put it there.

REFERENCES:

John M. Barry, *Rising Tide – The Great Flood of 1927 and How it Changed America*, New York, Simon & Schuster, 1997.

John Esten Cooke, *Robert E. Lee*, New York, G.W. Dillingham, 1893.

John F. Darby, Personal Recollection, St. Louis, G. I. Jones and Company, 1880.

Frederick J. Dobney, *River Engineers on the Middle Mississippi, A History of the St. Louis District, Army Corps of Engineers*, (U.S. Government Printing Office, Washington, D.C., n.d.).

Stella M. Drumm, "Robert E. Lee and the Improvement of the Mississippi River," *Missouri Historical Society Collections, 1928-1931*, Vol. 6.

Douglas Southhall Freeman, *R. E. Lee, A Biography*, 4, New York, Charles Scribner's Sons, 1935, I.

Gen. Charles Gratiot to J. B. Sarpy, July 1, 1837, autograph letter, signed.

Henry Kayser to Mayor William Carr, December 4, 1839, autograph letter, signed.

Robert E. Lee to John Mackay, June 27, 1838, autograph letter, signed.

Robert E. Lee to Henry Kayser, February 1, 1838, autograph letter, signed.

Robert E. Lee to Henry Kayser, March 9, 1838, autograph letter, signed.

Robert E. Lee to Henry Kayser, June 16, 1840, autograph letter, signed.

Robert E. Lee to Henry Kayser, January 15, 1844, autograph letter, signed.

Henry M. Shreve, *Documents Accompanying the Report of the Chief Engineer, Hudson, Ohio, Mississippi, Missouri and Arkansas River Improvements, 1840.*

Kristie C. Wolferman, *The Osage in Missouri*, Univ. of Missouri Press, Columbia, Missouri, 1997.

Come to Nashville in 2002
for
ASCE's 150th Birthday Party
and
Join the Tennessee Section
In Celebration of Our Two Centuries of Civil Engineering

Dan Barge, Jr., F.ASCE[1]

When compared with the thirteen states composing the original colonies, Tennessee's recorded history is relatively young. Although claimed by the Revolutionary State of North Carolina at the 1776 Declaration of Independence, Tennessee was in fact Indian territory.

True, there had been activities in this area by French fur traders and the British military during the French and Indian War from 1755 to 1762. The English built a fort at Loudon under an agreement with the Cherokees, who promptly reneged and massacred the entire garrison. France lost the war and, among other conditions, ceded to the British all claims to land east of the Mississippi River. In 1763 the British, in a futile attempt to placate the Indians, issued a proclamation that no whites could settle west of the Appalachian Mountains.

Notwithstanding British decrees, by the early 1770s, there were four communities in the western foothills of these mountains. Concurrently, there were many land speculators attempting to buy the cheap Indian land and resell it at a profit to incoming settlers. The most notorious of these was Richard Henderson of North Carolina who bought about 20 million acres for six wagonloads of goods said to be worth 10,000 English pounds.

Henderson then hired James Robertson and John Donelson to organize and lead a party of some 300 pioneers, men, and animals overland approximately 250 miles to the Great French Lick in 1779. In 1780 Donelson led a flotilla of about 40 rafts, loaded

----

[1]Chairman Emeritus, Barge, Waggoner, Sumner & Cannon, Inc., 162 Third Avenue North, Nashville, Tennessee 37201-1811—ASCE President 1987

with furniture, hardware, women, children, and provisions, about 850 miles down the Tennessee River to the Ohio River, then up the Ohio and Cumberland Rivers for 200+ miles to join the overland party—a four-month journey. The men had built a fort, cleared land, and planted a crop in anticipation of the arrival of the women and children.

For the next three decades, there were regular and frequent Indian attacks, but the settlers kept coming—clearing, planting, and building stands, villages, towns, and even cities. By 1784 the French Lick settlement was able to get the North Carolina legislature to establish it as a town called Nashville. Twelve years later, June 1, 1796, the North Carolina territory west of the Appalachian Mountains was approved by the United States of America as its 16th state—Tennessee.

The early settlers were entrepreneurs, harvesting timber and making lumber and brick—not just for local use, but also for export—floating their products down the rivers to New Orleans. James Robertson, one of Nashville's founders, discovered iron ore and built numerous iron furnaces and forges using local ore, stone, and coke. He sold his largest operation, Cumberland Furnace, to Montgomery Bell, Tennessee's first true industrialist. Bell later expanded operations by constructing a tunnel through a narrow ridge to permit flow of the Harpeth River, a fall of 29 feet, to power trip hammers for the production of wrought iron. This *Patterson Tunnel* was designated a National Civil Engineering Historic Landmark in 1981 and is considered the oldest such tunnel in the United States.

This furnace is said to have produced cannonballs for use by Andrew Jackson's Army in the Creek Indian Wars and his victory over the British at New Orleans in 1815. Another of its famous products was the *sugar kettle*, a giant cast-iron pot, 6 to 8 feet in diameter, used in Louisiana to boil sugar cane juice into sugar. Cumberland Furnace closed in 1938 after a colorful life of 150 years.

After Eli Whitney invented the cotton gin in 1793 and Andrew Jackson and others treatied with the Indians of West Tennessee, the expansion of the new state accelerated. Large plantations were developed, using slave labor to produce cotton for shipment to mills in New England and Europe. The center of population shifted westward, so the state's capital was moved in 1812 from its original location of Knoxville to Nashville. However, after the legislature had met in various buildings in Nashville, they decided to return to Knoxville in 1817, then to Murfreesboro for eight years. Only after the city had acquired a four-acre site on a hill in the center of the small city and offered it to the state as a building site did the legislature return to Nashville as the permanent site in 1826.

Even with a site, the legislature could not agree on the necessary actions to commence the design and construction of the new State Capitol for sixteen more years. In 1844 the Governor was permitted to name a commission "to superintend the construction of a statehouse, to direct the labor of the penitentiary to the erection of the same, and thereby save the people of the state from taxation."

The first duty of the Board of Commissioners, composed of five leading citizens representing the entire state, was to select an architect. They wanted to stage a competition for designs, but could not agree upon a procedure. They purchased a set of plans for the recently completed State Capitol of North Carolina at Raleigh and received unrequested designs from at least three other architects, two of whom had designed other State Capitols.

The record is not clear as to how the Commission learned of William Strickland, who was chosen as the architect. His name first appears in the Commission's minutes of June 14, 1844, as a prospect. He was invited to visit Nashville and present a plan for the Commission's review. If his plan was adopted, he would be paid $500 or, at his option, be hired to superintend the erection of the building at "a salary of $2,500 per annum."

This portrait of William Strickland was painted by Washington Cooper during the architect's years in Nashville. It now hangs in the Capitol near the east entrance.

Figure 1. William Strickland

William Strickland in 1844 was 55 years old and a well-known architect with many significant buildings to his credit, such as the Masonic Temple (Philadelphia 1810), the first U.S. Customs House (Philadelphia 1819), and the U.S. Mint

(Philadelphia 1824). He also consulted as a civil engineer on canals, railroads, breakwaters, dams and locks, ports, and suspension bridges.

His career is an amazing story. Born in 1788 in Navasink, New Jersey, he was the eldest son of a carpenter who found work with Benjamin Latrobe on the construction of the Bank of Philadelphia. In 1803, at age 14, young William was apprenticed to Latrobe to learn the fundamentals of engineering and architectural practice and to work on the design of the U.S. Capitol in Washington and on the Chesapeake and Delaware Canal.

The apprenticeship lasted slightly more than two years because Latrobe questioned Strickland's dedication to work. However, the young man's talent for drawing and painting apparently drew attention because he was soon hired to do playbills for the Park Theater in New York. He returned to Philadelphia in 1808 and designed the Masonic Hall as a Classic Gothic structure—his first independent architectural project. During the War of 1812, he was employed by the City of Philadelphia to help design and supervise the defenses of the city.

In the next decade, Strickland had many successful building and engineering projects—churches, temples, and university buildings—as well as drawings of naval encounters during the recent war. He prepared illustrations for books and novels and made murals for commercial ventures. But his claim to fame came when he won the 1818 design competition for the Second Bank of the United States over his mentor, Benjamin Latrobe, and many other prominent architects of the time. This building cost over half a million dollars and took six years to complete, bringing much attention to the young architect.

In 1824 he was chosen by the city to make repairs to Independence Hall and to plan other arrangements for the return visit of General Lafayette. He designed a series of triumphal arches along the General's route from the dock to Independence Hall, where the main arch was constructed. Lafayette's secretary was very complimentary of Strickland's work, which added further to his distinction, so he was invited to deliver a series of lectures on architecture at the Franklin Institute.

This fame probably accounts for his next commission. In 1825 he and one of his pupils, Samuel H. Kneass, were engaged by the Pennsylvania Society for the Promotion of Internal Improvements to travel to England and make a report on the British advances in the design, construction, and operation of "canals, roads, railroads, harbors, gas plants, iron smelting, and printing of calico, to collect books and pamphlets on these subjects, and to get models of machines."

Strickland and Kneass sailed in March 1825 and returned in December, spending all of their time in England, Ireland, and Scotland. They published *Report on Canals, Railways, and Roads* in 1826, which was well subscribed to by Congress, colleges, canal companies, and individuals (probably engineers) interested in internal improvements (infrastructure). Copies of this report still exist, so your author hopes to design a program featuring these drawings and sketches at *ASCE's 150th Birthday Meeting* in Nashville, Tennessee, in October 2002.

Strickland's career had its ups and downs for the next fifteen years. The ups included many federal buildings, including branch U.S. Mints at New Orleans and Charlotte and the Naval Asylum in Philadelphia. He also made several trips to Europe, touring England, France, Italy, and eastern Germany. In 1842 he was elected to membership in the Royal Institution of Civil Engineers in England. The downs centered mainly on several canal projects which failed, including the Cairo City and Canal Company in Illinois, at the confluence of the Mississippi and Ohio Rivers, which flooded before the levees were completed. All buildings, docks, and improvements built in the first two years were destroyed. He also met with Benjamin Wright (designated by ASCE in 1968 as "Father of American Civil Engineering") and about 40 other practitioners in Baltimore in February 1839 to form an "American Institution of Civil Engineering," later changed to American Society of Civil Engineers. This organization's attempt faltered, as did others in other cities (except Boston in 1848), so the world had to wait until 1852 for the American Society of Civil Engineers and Architects. No records have been found to ascertain William Strickland's interest in ASCE. Perhaps he was too concerned at this time with the completion of Tennessee's State Capitol and other Nashville projects, or perhaps age was catching up with him in his 64th year.

Back to Nashville and the Capitol project—in July 1845, the legislature approved the Building Commission's report to adopt Strickland's design and proceed with the construction under his supervision, using prison labor to quarry, hew, then transport and erect the stone. The cost was calculated by Strickland at $340,000, which might be reduced by $85,000 to $100,000 by employment of prison labor.

Strickland's design for the Capitol is described by historic architects as the epitome of Classic Greek Revival Architecture. It is 238 feet by 109 feet with three stories (and useful basement), centered entrances on all four sides, and eight columns surrounding the entrances, all of which have substantial alcoves. It has a slim, square-based Grecian tower, topped by an octagonal shaft and a round, fluted dome. The tower straddles wrought iron trusses spanning the 109-foot width. Incidentally, the wrought iron for the trusses and tower was manufactured at Cumberland Furnace, some 40 miles west of Nashville (see page 2). The interior, with its tall ceilings and wide stairways with lavish ironwork, reflects superior craftsmanship.

To say that the construction went well would be a terrible overstatement. First, the Commission was beset by regional jealousy and politics. The party of Andrew Jackson was fading. James K. Polk was elected President of the United States in 1844, but did not carry his home state of Tennessee. Then came the Mexican War, during which Tennessee earned the reputation of "Volunteer State," which further divided the Commission's attention. The use of prisoners as workers in the building activities was also stopped, and the legislature played games with the appropriations.

Figure 2. Tennessee State Capitol Following 1956 Restoration

By October 1849, more than four years after the laying of the cornerstone, the cut stone bearing walls were barely above first-floor level, and half of Strickland's estimated cost had been spent. Everyone blamed everyone and everything else, including a cholera epidemic at the prison quarry. But a new Governor demanded action with new appropriations, firm orders for the iron materials and marble, and changes in the makeup of the Building Commission. Good results were had and the building was roofed and sealed in by early 1852 so that interior finish work could commence.

Finally in October 1853, following Andrew Johnson's inauguration as Tennessee's Governor, the legislature met for the first time in the new Capitol. It was not nearly complete, however, and required another six years to finish all building work.

William Strickland, who had moved to Nashville in 1845 and lived with his family at the City Hotel, died suddenly in April 1854 and was interred in a crypt at the northeast portico of the Capitol. His son, Francis, who had worked for several years on this project, was engaged to complete his father's assignment. Francis was discharged after two years, following squabbles with the Commission about finish details and cost overruns.

This Capitol project was declared complete, with the exception of the grounds, in March 1859 at a cost of $880,000. It is paradoxical that the grounds work was still not complete in early 1862 when Andrew Johnson, by then Military Governor of Tennessee, took over the new Capitol and fortified it by placing artillery behind barricades on the porticos and domiciling Union soldiers in tent camps on all sides of the grounds. Incidentally, Governor Johnson was elected Vice President of the United States in 1864, and became President when Abraham Lincoln was assassinated in April 1865.

In 1960 a major restoration of the building was necessary, replacing all exterior cut limestone walls and columns with a more durable Indiana limestone. The structure is now well maintained and attractively appointed. The Tennessee Section of ASCE plans to nominate the Capitol for designation as a Historic Landmark in 2002. The unveiling of the plaque would be one of the Annual Convention's highlights.

There are several more interesting paragraphs in the development of civil engineering in our state, but these will have to wait for the *great party* at ASCE's sesquicentennial anniversary.

The dates and facts used in this paper come from many sources, among which are Tennessee Blue Book (1796-1996); American Civil Engineer (1852-1974), Wm. H. Wisely; A Biographical Dictionary of American Civil Engineers, ASCE; Our Ancestors Were Engineers, A. W. Crouch and H. D. Claybrook, published by Nashville Section ASCE 1976; Tennessee State Capitol, Tennessee Historical Quarterly, Fall 1966; and William Strickland Architect and Engineer, Agnes A. Gilchrist, 1950.

HISTORY AND HERITAGE PROGRAM FOR THE MARYLAND SECTION

Michael A. Ports[1], PE, F. ASCE and Steve E. Yochum[2], A. M. ASCE

Introduction

The State of Maryland is endowed with many important civil engineering landmarks, structures, systems, and facilities with considerable national and local historic significance. Unfortunately, to all but a few, most are well kept secrets. Even if identified as such, obtaining information about them is difficult and time consuming. To help to remedy the situation, the Maryland Section established a History and Heritage Committee.

The committee was established late in 1993 under the present leadership. Volunteers are solicited continuously from the membership. At present, the committee has two primary program activities. The first is the State History and Heritage Landmark Award Program. The second is the preparation of the History and Heritage of Civil Engineering in Maryland Booklet.

State History & Heritage Landmark Award Program

The committee solicits nominations for Landmark Award recognition from the membership through meeting and newsletter announcements. In addition, the committee works to develop and maintain a short list of ten to a dozen suitable candidates. During the year, the committee carefully considers each candidate for selection. The winner is recognized and asked to prepare a short technical presentation for the Section's Annual Awards Banquet in May.

---

[1] Principal Professional Associate, Parsons Brinckerhoff Quade & Douglas, Inc., 301 N. Charles Street, Suite 200, Baltimore, Maryland 21201
[2] Water Resources Engineer, Parsons Brinckerhoff Quade & Douglas, Inc., 301 N. Charles Street, Suite 200, Baltimore, Maryland 21201

The historical relevance of the project to both society and the civil engineering profession are considered in the selection process. Of course, nominated projects may relate to any discipline within the field of civil engineering.

To date, the selected winners include the following:

1. The Foxcatcher Farm Covered Bridge was the first State Landmark Award recipient in 1994. The structure is located on Big Elk Creek at the Fairhill Natural Resource Area. Designed and built by Ferdinand Wood in 1860, the structure was completely restored in 1992. The bridge established a design which quickly became the design standard for that time within Maryland and much of the country. It is the oldest of only four remaining Burr Arch bridges in Maryland and the sole remaining Burr Arch truss carrying vehicular traffic. Three nationally recognized bridges share similar attributes, including the Blenheim, Bridgeport, and Cornish-Windsor Covered Bridges.

2. The Jerusalem Mill, located in the Gunpowder Falls State Park in Harford County, was recognized in 1995. Built in 1772 by David Lee, the grist mill grew into one of the largest mills in Maryland and contributed to making the Baltimore area one of the largest and most important milling centers in the country. The mill race channels water under the mill centers in the country. The mill race channels water under the mill through two segmented arches in the stone walls. This rare design is known as a vertical shaft mill, in which a horizontal water wheel turns an upright shaft which powers the machinery above. The mill represents an important part of the economic history of the region. Restoration was completed in 1994. The structure serves as a museum and learning center for the State Park.

3. The Lake Roland Dam and Gatehouse was recognized in 1996. The facility was designed by James Slade as part of a three reservoir system to supply Baltimore City with potable water. Construction began in 1858 and completed by 1860. The 500 million gallon capacity was held in check by a dam 60 feet thick and 40 feet from base to crest. The 125 foot spillway runs along the top of the rubble core structure with wing walls enclosing earth embankments rising six feet above the crest. On the parapet next to the southern wall, Charles P. Manning erected a Greek revival gatehouse, whose floor made a platform protecting a system of tunnels through which water flow was controlled. The facility functioned as an integral part of the water supply

system for more than fifty years and was completely restored in 1994.

4. In 1997 the Trolley Station at Old Bay Shore Park was recognized. Built in 1906, in conjunction with an amusement park, the trolley station made Bay Shore Park accessible to many Marylanders. The original building, which resembles a large pavilion, is 208 feet long, 48 feet wide, and constructed of Georgia pine timbers. The roof consists of exposed timber trusses with full length lumber. Much of Bay Shore Park's commercial success is attributed to the efficiency of the trolleys. The United Railways & Electric Company, which also owned the park, ran a double track line with comfortable cars seating 50 passengers. The trolleys transported up to 1,000 passengers per hour for a round trip fare of just 30 cents.

5. The latest recipient was the Chesapeake & Ohio Canal, awarded in 1998. The 184.5 mile long canal rises from near sea level at Georgetown to 605 feet in elevation at Cumberland, Maryland. The canal consists of 11 aqueducts, 74 lift locks, 65 culverts, 12 feeder locks, 1 lift bridge, and 1 tunnel. The canal was conceived by George Washington and later implemented by the C&O Canal Company, with financial backing from the State of Maryland. Ground was first broken for construction in 1828. By 1850, the canal was completed to Cumberland and connected our Nation's capital to the Potomac River Valleys vital resources. With a top width of 40 to 50 feet and a 12 foot wide tow path, the canal required an incredible amount of earth moving considering the tools of the day.

A complete file for each awardee is prepared and archived. A copy of the file is provided to the State Historic Preservation Officer and to the Maryland Historical Society Library. A total of more than one dozen projects is on the current short list. Because the short list is so long, multiple awards are planned. For the next few years, the February or March Section meeting will be dedicated to recognizing multiple History and Heritage Landmark Awardees.

History and Heritage of Civil Engineering in Maryland Booklet

Earlier this year, the Maryland Section was awarded a national ASCE State Government Relations grant to prepare and publish a booklet entitled *"History and Heritage of Civil Engineering in Maryland."* The small booklet (8.5" x 5.5") will contain two page descriptions and graphics about fifty significant projects around the State. The booklet also will contain a map of the State indicating the location and access to the fifty sites. In addition, the electronic equivalent of the booklet will be added to the Section web site.

A short list of 60 to 70 candidate projects will be selected from:

- Section Outstanding Civil Engineering Achievement Awardees.
- Section History and Heritage Landmark Awardees.
- Engineering landmarks on the National Register of Historic Places.
- National History and Heritage Landmark Awardees.
- Information from local museums and historical societies.
- Suggestions from membership.

Two civil engineering undergraduate students are now gathering the required information and preparing the draft document. The booklet is scheduled for final draft and publication in September 1998 and will showcase the most significant civil engineering projects in Maryland.

Conclusion

The activities of the committee are intended to increase the awareness of civil engineering. The benefits include:

- Increase public awareness of the role of civil engineering in shaping the history of the country.
- Provide resources which capture the importance of the unique infrastructure that is often taken for granted.
- Provide tools to help educate younger students considering civil engineering careers.

A GUIDE TO CIVIL ENGINEERING PROJECTS IN AND AROUND NEW YORK CITY

Robert A. Olmsted[1], P.E., F.ASCE
and Michael N. Salgo[2], P.E., Hon. M.ASCE

ABSTRACT

Many of the world's foremost civil engineering projects -- bridges, tunnels, highways, skyscrapers and buildings, airports, water supply and environmental systems -- are located in New York City and its environs. In order to provide visiting engineers and the public with an overview of some of the major projects in and near the city, the ASCE Metropolitan Section published a 36-page illustrated booklet entitled, *A Guide to Civil Engineering Projects in and around New York City* in 1997. The guide, which describes over 52 projects, gives the general public a better understanding of what civil engineers do and build. Several projects are ASCE Historic Civil Engineering Landmarks, National Historic Landmarks, or are listed on the National Register of Historic Places.

INTRODUCTION

The histories of American civil engineering and the City of New York are closely linked. The opening of the Erie Canal in 1825, whose Chief Engineer, Benjamin Wright, ASCE calls "the father of American civil engineering", sparked the rapid growth of the city and its need for massive infrastructure: transportation, water supply, buildings, etc. An early need was to develop an adequate water supply system. Following a disastrous fire and cholera epidemic in the 1830s, the City built the Croton water supply system. That system, with its upstate dams and reservoirs, a 66 km (41 mile) long aqueduct, and the 445 m (1,460 ft) long High Bridge with its 15 stone arches, the nation's longest bridge when built, was the

----

[1] Transportation Consultant, Chair, ASCE Met Section History and Heritage Committee, 33-04 91st Street, Jackson Heights, N.Y. 11372-1752
[2] Chair, ASCE Met Section Hospitality Committee, 137-32 76th Avenue, Flushing, N.Y. 11367-2818

most outstanding municipal water supply works at the time and the prototype for later projects. Appropriately, it was in the office of the Croton Aqueduct Department that the **American Society of Civil Engineers** was founded in 1852.

THE GUIDE

In 1995, the Met Section's International Group, then chaired by Bruce Podwell (later by Nick Massand), decided that it would be a good idea to publish a guide to acquaint foreign visitors with engineering projects in New York City. The guide committee was expanded to include the Hospitality Committee and the History and Heritage Committee. Because the History and Heritage Committee had the technical facts and historical background, the guide's text was largely prepared by the History and Heritage Committee. The scope was broadened to produce a document not only for visiting engineers, but for the general public as well. Thus the final *Guide* promotes greater public awareness of the civil engineering profession and what civil engineers do. Finally published in 1997, the *Guide* is a 36-page illustrated booklet designed to slip in one's pocket, or a No. 10 envelope for easy mailing. Over 52 projects are described. Wherever possible, a project's engineer is identified.

Many of New York's engineering wonders can be seen by taking a sightseeing boat ride around Manhattan Island, and the first 23 major projects are described as seen while sailing around Manhattan in a counterclockwise direction.

Our tour begins at the Statue of Liberty, an International Historic Civil Engineering landmark built in 1886. It is well known that Frederic Auguste Bartholdi was the sculptor. Lesser known is that the statue is composed of a thin copper skin attached to a structural iron frame (armature) that was designed by Gustave Eiffel, who later built the Eiffel Tower (an early example of curtain wall construction). The statue stands on a stone and concrete pedestal which was the largest single mass of concrete poured at the time. One of the wonders of the world, the statue was extensively rehabilitated in 1986.

Sailing up the East River, we see the ventilation buildings of Ole Singstad's Brooklyn-Battery Tunnel (1950), the longest underwater vehicular tunnel in North America at 2780 m (9,117 ft). Then comes John and Washington Roebling's Brooklyn Bridge (1883), a National Historic Civil Engineering Landmark often called the eighth wonder of the world, its 486.4 m (1,595 ft 6 in) span shattered records. The next two bridges are the Manhattan (1909) and Williamsburg (1903). The Williamsburg Bridge's designer, Leffert Lefferts Buck, was charged to build a bridge longer than the Brooklyn Bridge -- it's suspended span is 1.4 m (4.5 ft) longer) -- and for less money (a source of some of today's problems). The first major bridge to use steel instead of masonry for the towers, the "Willie B", was the world's longest for 23 years. It is currently undergoing a major renovation. The approach spans are being completely replaced, and the roadway is being replaced with an orthotropic deck. Leon Moisseff's Manhattan Bridge is a two-level suspension bridge which carries four heavy-duty subway tracks. The torsional force caused by unbalanced subway train loadings has been a major problem. This 448 m (1,470 ft) bridge is also undergoing major renovation.

Between the Manhattan and Williamsburg Bridges lies the Brooklyn Navy Yard -- the "can do yard". Now an industrial park, many of the nation's most famous warships were built here: the battleship Maine, sunk in the Spanish American War; the battleship Arizona, sunk at Pearl Harbor, and the battleship Missouri, on whose decks World War II ended in Tokyo Bay in 1945.

Continuing our sail, we see the United Nations Headquarters, the Franklin D. Roosevelt (East River) Drive, the Queensboro Bridge (1909), and the expanded New York Hospital. The FDR Drive is part of an express highway ringing Manhattan Island. The most spectacular sections are the double-deck sections straddled by buildings and esplanades. A section of the drive, dubbed "Bristol Basin", was built on fill using rubble from bombed English cities brought over as ballast in wartime ships. New York Hospital has just been expanded on a platform built over the Drive, a difficult construction project that won the ASCE Met Section Construction Achievement Award for 1997. The Queensboro Bridge, a multi-span cantilever bridge designed by Gustav Lindenthal, is also undergoing major renovation. Just north of the bridge is the Roosevelt Island Tramway, the only aerial tramway performing regular public transit service. Paradoxically, the four East River bridges, which were designed in the pre-automobile age, carried more people in their early years. As street car and elevated railway tracks were abandoned and replaced with automobile roadways, the people-carrying capacity of the bridges was reduced.

The Triborough Bridge (1936), a National Historic Civil Engineering Landmark, hovers into view. An achievement of Commissioner Robert Moses and Engineer Othmar Ammann, it was the first major bridge project to incorporate extensive highway approaches as an integral part of the project. The 5.6 km (3.5) mile long structure consists of three major bridges and connecting viaducts: an 830 m (2,724 ft) suspension span; a 95 m (310 ft) vertical lift span (world's largest when built); and a 107 m (350 ft) fixed truss. Parallel to the suspension bridge is Gustav Lindenthal's impressive 298 m (977 ft 6 in) Hell Gate Bridge (1917), a four-track railroad bridge which was the longest and heaviest steel arch in the world when built. It is now an integral link in Amtrak's Northeast Corridor high speed rail line between Washington and Boston.

We enter the Harlem River which separates Manhattan (left bank) from The Bronx (right bank). Fourteen bridges span the Harlem; seven are center-pier, swing bridges (possibly the world's largest collection); two are vertical lift bridges; three are steel deck-arches; and one is a combination masonry arch viaduct and steel arch bridge. Many of these bridges were built after 1895 to meet new navigational requirements. Of the swing bridges, the Macombs Dam Bridge (1895) was the heaviest movable bridge in the world when built, while the University Heights Bridge was originally built at another location in 1895 and floated to its present location in 1908. The High Bridge (1848), the oldest extant bridge in New York City, was built to carry the Croton Aqueduct across the Harlem River. Designed by John B. Jervis, it originally consisted of 15 Romanesque stone arches. In 1927, five of the stone arches were replaced with a 98 m (322 ft) steel arch to increase navigational clearances. (The Croton Water Supply System is a National Historic Civil Engineering Landmark.) The 256 m (840 ft) arch of the Henry Hudson Bridge (1936), which carries the Henry Hudson Parkway over the Harlem, was the longest steel plate-girder fixed arch bridge when built. Interestingly, as a young student, David B. Steinman designed a bridge at this location for his dissertation; years later, he fulfilled his dream as one of its designers.

As we turn south into the Hudson River, Othmar Ammann's masterpiece, the majestic, 14-lane George Washington Bridge, comes into view. Its 1067 m (3,500 ft) suspension span doubled the previous record. A National Historic Civil Engineering Landmark, the impressive steel towers were originally planned to be clad with masonry. Fortunately, this was not done due to lack of money. A lower deck was added in 1962.

The North River Water Pollution Control Plant, one of the world's largest waste water treatment plants, is built on a platform extending over the Hudson River (also known as the North River). The 11 hectare (28 acre) concrete structure is supported on 2,800 caissons pinned into bedrock. The plant, which serves a daytime population up to 1,000,000 persons, is covered by a public park on its roof.

Continuing south, several tunnels are crossed. The three-tube Lincoln Tunnel (1937) is the world's busiest vehicular tunnel. Amtrak's North River tunnels (1910) are an integral link in the busy northeast corridor rail system. The uptown PATH subway tunnel, which was begun in 1874 but not completed until 1908, was the first attempt to build an underwater tunnel and the first to use a shield in the United States. It is a National Historic Civil Engineering Landmark. The Holland Tunnel (1927) was the first long, mechanically ventilated vehicular tunnel in the world, A National Historic Civil and Mechanical Engineering Landmark, the Holland Tunnel is named after its designer, Clifford M. Holland, who died during the course of construction, one of the few engineering works named after its engineer. By the time we return to the Statue of Liberty, the boat will have crossed over 17 subway and railroad tunnels (41 tracks in all) and four vehicular tunnels, and under 20 bridges.

New York City is famous for its skyscrapers. Although a Chicago development, the tallest building in the world was in New York for most of the 20th century. Three engineering developments made skyscrapers possible: the elevator, steel frame construction and rapid transit. Almost lost among its neighbors, and better known these days for its major tenant, a well known music and computer store, the 118 m (386 ft) high Park Row Building (1899) was the world's tallest for nearly a decade. Later, the distinctive campanile clock tower of the 213 m (700 ft) high Metropolitan Life Tower (1909) dominated the skyline for five years. The most famous of these early skyscrapers was the 241 m (792 ft) high Woolworth Building (1913). Paid for in cash with nickels and dimes collected by Frank Woolworth at his five and ten cent stores, this Gothic style, terra-cotta clad, landmark reigned supreme for 17 years.

In the pre-depression roaring 20s, there was a frantic race to build the tallest building. Forty Wall Street thought it had won that race with its needle-like spire. (Fixed spires count, but not flagpoles or antennas.) But another building secretly fabricated a spire in a shaft and thus, after the surreptitious pinnacle was raised, the 320 m (1,048 ft) high Chrysler Building (1930) won the race. But not for long. A few blocks away the quintessential skyscraper, the Empire State Building (1931), was nearing completion. The 102-story Empire State Building was built in record time. Its 58,000 ton steel frame was erected in less than six months! The topmost 61 m (200 ft) was designed as a mooring mast for dirigibles, an idea found to be impractical. The building, which ASCE has named one of the seven modern civil engineering wonders in the United States, survived a direct crash of an Army bomber in 1945. The Empire State Building was king for 41 years until the 110-story, 417 m (1,368 ft) high twin towers of the World Trade Center opened. But now the tallest buildings are in Chicago and Asia.

New York City could not function without its vast public transportation system. Although the first American subway was built in Boston a hundred years ago, New York's is the largest. The original IRT (Interborough Rapid Transit) subway, which opened between 1904 and 1908, was the world's first four-track subway (for local and express service). The route included two of the earliest subaqueous passenger tunnels and the second longest rock tunnel at the time (only Massachusetts' Hoosac tunnel was longer). The first subway, whose chief engineer was William Barclay Parsons, is a National Historic Civil and Mechanical Engineering Landmark. A vast expansion, which was a public works project comparable to the Panama Canal in scope, was largely built between 1910 and 1920 and resulted in two subway systems (IRT and BMT). This "dual system" included one of the world's first immersed tube subaqueous tunnels (Lexington Avenue/Harlem River). A third system, the "independent (IND) subway system", was built in the 1930s. In 1940, the three systems were unified as one municipally operated system. Today, the New York subway comprises 390 km (230 miles) of route, [220 km (137 miles) underground], 1055 km (656 miles) of running track, 468 stations and 14 underwater tunnels. Nearly 6,000 subway cars carry 3.8 million passengers each weekday.

Two major railroad stations serve New York City: Penn Station and Grand Central Terminal. The Hudson (North) River was a formidable barrier for the railroads approaching the city from the west. Between 1900 and 1910, the Pennsylvania Railroad built two single-track, shield-driven, tunnels under the Hudson River, four similar tunnels under the East River, the 21-track Penn Station, a major storage yard in Sunnyside, Queens, and associated work. The privately-funded project would cost over $6 billion to replicate at today's costs. The outcry following the 1962 demolition of the station building, an architectural masterpiece, led to New York City's Landmarks Preservation Law, one of the first in the country. The below-ground station still operates and is being extensively modernized. Penn Station, which is owned by Amtrak, is a major commuter terminal and the busiest railroad station in the nation. Not to be outdone, the Pennsylvania Railroad's arch rival, the New York Central Railroad, built its majestic Beaux Arts Grand Central Terminal in 1913 (replacing an earlier station). Electrification made it possible to build this two-level station and construct large office buildings over the tracks, creating the prime Grand Central business district. The outstanding feature of the terminal is the 38 m (125 ft) high main concourse and its recently cleaned arched constellation-painted ceiling. Grand Central Terminal is currently being renovated.

We saw some of New York's great bridges on our sightseeing boat ride. But there are many more. In fact, there are nearly 2,100 bridges of all types in the city, 76 of them over waterways. The 27-year period from 1883 to 1910 was the greatest era of waterway bridge building the world has ever seen when 23 bridges were built, ranging from the unique Carroll Street retractile bridge to the Brooklyn Bridge. Recent large bridges include Othmar Ammann's Verrazano-Narrows (1964) and Bayonne (1931) bridges. The 1299 m (4,260 ft) suspended span of the 12-lane, two-level, Verrazano-Narrows was the longest in the world for 19 years. The 510 m (1,675 ft) steel arch of the Bayonne Bridge, a National Historic Civil Engineering Landmark, was the longest arch in the world for 46 years. The 498 m (1,632 ft) suspension span of the upstate Bear Mountain Bridge (1924), a Local Historic Civil Engineering Landmark, held the title of the world's longest for two years. And a bit of trivia. New York's outermost bridge, a 229 m (750 ft) cantilever, is the Outerbridge Crossing. Why? It is named after Eugenius H. Outerbridge, first chairman of the Port Authority of New York and New Jersey!

These bridges and tunnels are connected by a well developed highway system: parkways for passenger car traffic, and expressways for mixed traffic. The New York region pioneered the development of limited access highways. The world's first grade separated roadways were built in Central Park in the 1860s. The term parkway was coined by Frederic Law Olmsted in 1868 to describe tree-lined boulevards connecting parks and several were built in the late 1800s. The world's first grade separated, limited access highway of any considerable length was the 80 km (50 mi) Vanderbilt Motor Parkway on Long Island, which was privately built about 1910 for automobile races. The first modern parkway was the Bronx River

Parkway (1924), which became the model for similar highways throughout the world. Between the 1930s and 1960s, during the heyday of Robert Moses, hundreds of miles of modern, limited access parkways and freeways (called expressways in New York), were built.

Parks are not often thought of as engineering works. But New York City's Central Park (1858), which was the first major landscaped park in America, was a major engineering achievement. Designed by Frederick Law Olmsted and Calvert Vaux, nearly 3.8 million cu m (5 million cubic yards) of earth and rock were moved to reshape the barren landscape. Many miles of drain tile were installed, new ponds and lakes were created, and driveways were built. Some 40 stone and iron bridges were built, including the 26.5 m (87 ft) long wrought and cast iron Bow Bridge, one of the oldest iron bridges in the United States. A major innovation was the construction of the first grade-separated roads in the country. Constructing these four sunken transverse roads required blasting deep cuts and a tunnel through solid rock

The Guide describes the New York region's three major airports: LaGuardia, Newark (N.J.) and Kennedy airports. Newark International Airport (EWR), the region's oldest airport (1928) and one of the first to have a hard-surfaced runway, is a National Historic Civil Engineering Landmark. The airport has been expanded over the years from its original 27 ha (68 acres) to 820 ha (2,027 acres). A new monorail connecting the terminal buildings and parking lots opened in 1996. John F. Kennedy International Airport (JFK), which opened in 1948, was built on hydraulic fill pumped out of Jamaica Bay. The longest runway is 4443 m (14,572 ft). An elevated automated light rail system is about to be built connecting the terminals with each other, remote parking lots, and nearby subway and railroad stations. LaGuardia Airport (LGA) (1939), the smallest airport of the three, was built partly on filled land reclaimed from the East River. The two original runways have been lengthened to 2134 m (7,000 ft) on pile-supported concrete platforms extending over water. New and enlarged terminals have been built over the last few years, and a direct rail connection to the City is under consideration.

So far, transportation projects have dominated the Guide. While New York City could not function without its bridges, tunnels, highways, subways, railroads and airports, neither could it exist without its vast environmental infrastructure. As the city's population grew by leaps and bounds in the 1800s, the landmark Croton water supply system became inadequate. Between 1885 and 1906, the Croton system was enlarged. The new masonry Croton Dam (1906) was the highest dam in the world. Between 1907 and 1926, a new system was built 160 km (100 mi) north of the city in the Catskill mountains. The 148 km (92 mi) long Catskill aqueduct crosses the Hudson River in a rock tunnel 340 m (1,114 ft) below the river. The newest system, which was built between 1937 and 1965, taps the headwaters of the Delaware River. Four major reservoirs were built in the western Catskills. Water is conveyed to the city through the Delaware Aqueduct, a 136 km

(85 mile) long pressurized rock tunnel. The dependable yield of the system is 1,290 mgd. Water is distributed through the city through three large water tunnels. The 113 km (70 mile) long third tunnel, still under construction, is one of the largest public works projects in the country. Wastewater is treated at fourteen water pollution control plants, of which the North River plant is the largest.

ASCE cooperates with the Library of Congress and the National Park Service in the Historic American Engineering Record (HAER) program which documents historically important structures in detail. New York City projects which have been documented by HAER are: the Roeblings' Brooklyn Bridge, John B. Jervis's Croton Water Supply System, William Barclay Parsons' First New York (IRT) Subway, Clifford Holland's Holland Tunnel, Othmar Ammann's George Washington Bridge, and Ammann's Bayonne Bridge.

These projects were designed and built by civil engineers who were leaders in the profession. Many were active in ASCE and served the society in top positions, including president. Met Section engineers look forward to building the next century's infrastructure.

CONCLUDING REMARKS

The Guide has been well received, and perhaps some day a larger second edition will be written. The authors wish to thank the other members of the Guide Committee for their tireless efforts: Armen Boyajian, Simon Fridman, Dan Garvey, Christian Ingerslev, Chee K. Lai, and Genaro Lozano.

REFERENCES

Metropolitan Section, American Society of Civil Engineers (1997). *A Guide to Civil Engineering Projects in and around New York City*, New York, N.Y.

Schodek, Daniel L (1987). *Landmarks in American Civil Engineering*. MIT Press, Cambridge, Mass.

Historic American Engineering Record (HAER). (1979). *Interborough Rapid Transit Subway*. Washington, D.C.

ASCE and the C&O Canal

Bernie Dennis[1], Bruce Mattheiss, P.E.[2], Steve Pennington, P.E.[3]
ASCE National Capital Section

On July 4, 1828, President John Quincy Adams symbolically "broke ground" for the Chesapeake and Ohio Canal. By 1850, construction of the canal was complete to Cumberland, MD. The 184.5-mile long canal featured 74 locks with 8-ft lifts; 7 diversion dams; a long tunnel (3118 feet) and a series of masonry aqueducts, major and minor, crossing the many tributaries of the Potomac.
--From the book "**Civil Engineering Landmarks of the Nation's Capital**" 1982, by the History and Heritage Committee of the National Capital Section, American Society of Civil Engineers.

BACKGROUND

The Chesapeake and Ohio (C&O) Canal is a national treasure and a symbol of civil engineering achievement. Situated along the north shore of the Potomac River, it follows the river from Georgetown in the District of Columbia to Cumberland, Maryland, a distance of 296 km (184 miles). Through the efforts, 40 years ago, of Supreme Court Justice William O. Douglas and others, the C&O Canal was designated a National Historic Park and placed under the care of the National Park Service (NPS) of the Department of the Interior. In 1982, the C&O Canal was designated a Local Historical Civil Engineering Landmark by ASCE's National Capital Section (NCS). In 1998, the C&O Canal was designated as a Local Historical Civil Engineering Landmark by ASCE's Maryland Section.

Civil engineers of the NCS are taking an active role in assisting the National Park Service in efforts to restore and preserve the canal. Recent NCS efforts began in

[1] Structural Engineer, Delon Hampton & Associates, 800 K Street, Suite 720 - North Lobby, Washington, DC 20001
[2] Resident Engineer, CH2M Hill, 625 Herndon Parkway, Herndon, Virginia 22070
[3] Director, Instrumentation Division, GeoServices, Inc., 8001 Cryden Way, Forestville, Maryland 20747

December 1995 with a response to a request from the C&O Canal Association for help in saving the Monocacy Aqueduct. As a result of early efforts on the aqueduct, the NPS realized that ASCE can and should play a role in restoring the canal and its historic structures. To this end, ASCE was invited to send representatives to a three day Sustainability Workshop in June 1996.

Subsequently, the NCS, the C&O Canal Association, and the NPS joined in a partnership to focus attention and resources on the stabilization and restoration of the Monocacy Aqueduct. In this partnership, the NCS offered assistance in three areas: financial, publicity, and technical support.

This paper summarizes ongoing efforts of ASCE/NCS to meet its commitment as a working partner in the preservation of the historic aqueduct structure.

Financial Commitment

The C&O Canal Association assists the NPS by directing the C&O Canal National Park Foundation. In the spring of 1996, the Association established the Monocacy Aqueduct Fund within the foundation, to collect funds specifically earmarked for the aqueduct.

To spark the fundraising effort, the NCS Board of Directors approved a contribution of $2,500 to the C&O Canal Association's Monocacy Aqueduct Restoration Campaign. The NCS contribution provided an incentive for further funding from the private sector and for matching funds from the NPS. On April 18, 1996, the NCS and C&O Canal Association presented a combined check for $10,000 to Mr. Bruce Babbitt, Secretary of the Department of the Interior, for restoration of the Monocacy Aqueduct.

Since that initial effort, the Monocacy Aqueduct Fund has grown to over $25,000. Through ASCE's Public Relations Grants Program, the NCS has obligated an additional $5,000 to be applied to the purchase of instruments as part of an ongoing monitoring program for the aqueduct.

Publicity Commitment

Publicity for the aqueduct, within the engineering community, was immediately provided through the *NCS Newsletter* going out to its 3300 members. Broader publicity was provided during ASCE's 1997 National Convention in Washington, DC, where the C&O Canal Association had a poster presentation addressing efforts to save the aqueduct. National publicity has been provided through articles in *"ASCE News"* and several trade journals.

Members of the NCS History and Heritage Committee participated in several of the C&O Canal Association-sponsored annual hikes that terminated at the Monocacy Aqueduct: the Douglas Hike in 1996 and the Heritage Hike in 1997. During both of these events, NCS set up and staffed a table with information about ASCE and its efforts to assist in the engineering studies of the aqueduct. Members were available to answer questions on any number of civil engineering subjects and allowed visitors to examine the aqueduct closely through surveying instruments. Following the hike in 1997, NCS delivered the keynote speech at the C&O Canal Association annual dinner addressing the ongoing engineering activities being provided by the Section in support of Park Service programs at the aqueduct.

NCS members also provide tours of the aqueduct to interested groups, potential benefactors, corporate representatives and distinguished federal and state political figures. ASCE is also interacting with organizations such as the National Trust for Historic Preservation and the Association for Preservation Technology, working toward the common goal of preservation awareness.

Technical Support

As part of its efforts to rebuild the canal following the devastating destruction of the January 1996 flood, the NPS addressed issues of sustainability in the design of repairs to help minimize such damage in future floods. To help define sustainability measures they conducted a three-day Sustainability Workshop. The NPS invited technical representatives from the US Army Corps of Engineers, NPS's Denver office, C&O Canal maintenance staff, and ASCE to participate and contribute ideas during the workshop, which included visits to key sites damaged during the flood. In response to the NPS invitation, NCS sent a team of four engineer members with backgrounds in hydrology, climatology, flood plain management, and historic canal structures.

Restoration of historic structures along the canal is another area of interest in the NPS. The restoration of the Monocacy Aqueduct is one of the highest priorities of the Canal Superintendent, specifically since it is the largest of the aqueducts on the canal and is key to the continuity of the canal. The Superintendent of the Canal outlined his vision of ASCE's role as a technical advisor in the NPS. In 1997, as the NPS prepared a Scope of Work for a condition study of the aqueduct, ASCE/NCS members assisted by providing constructive technical input to define study parameters and goals for the final report. ASCE/NCS members participated in the review of the 50% contract submittal in April 1998, and will play a key role in the review of resulting reports and design documents, later in 1998.

In addition, as continued technical support to the NPS, members of the History and Heritage Committee of ASCE/NCS proposed a long term performance monitoring program to collect data on the structure's behavior as it responds to the surrounding environment. The NPS accepted this proposal and the program is currently underway.

BEHAVIOR STUDIES FOR THE MONOCACY AQUEDUCT

The Monocacy Aqueduct is the largest engineering structure within the C&O Canal National Historic Park system. It is a grand, historic structure and in need of immediate help. Built in 1833, the aqueduct is 156 m (512 feet) long with seven masonry arch spans (see Photo 1). The aqueduct is the largest of 11 stone masonry aqueducts on the C&O Canal. The structure spans the Monocacy River between Frederick and Montgomery Counties, near Dickerson, Maryland.

The structure continues to stand despite numerous floods and attempts by Confederate soldiers to destroy it during the Civil War. In 1972, the flood waters of Hurricane Agnes ripped away major portions of the cast iron railing on the structure, tearing out many of the capstones on the downstream side of the structure. In response, the NPS and the Federal Highway Administration installed a steel harness to help temporarily stabilize the structure until resources and funds could be found to restore it. Since that time, little has been done to maintain the structure and the harness. Most recently, the flood waters in January 1996 crested over 6 feet above the top of the structure which normally towers 7.6 m (25 feet) above the Monocacy River. In August 1996 flood waters again rose above the structure.

The ASCE/NCS performance monitoring program is designed to gather continuous data regarding the structure's response to its environment. Changes in horizontal and vertical alignment, the tendency for overturning, movement of cracks in the masonry matrix as well as joints between stones, arch ring and spandrel wall deflections are among the types of data being obtained (see Photo 2). We are also gathering data on the condition of the steel harness: determining the existing tension in the tie rods and will soon install strain gages to monitor the response of the harness during flood conditions. As the program evolves an automated data logging system will be installed. This logging system will be powered by solar cells and communication provided with a cellular telephone link. The overall plan will be for both the Park Service and NCS to have computer linkups with the onsite instruments to give a continuous readout of the structures behavior as environmental changes occur.

In addition to structure data, environmental data will be continuously monitored as well. Temperature, pressure and rainfall data will be logged simultaneously with structure data. River data will also be incorporated to include stage heights and velocity conditions. This information will also be part of the logging system and be transmitted along with performance data. The ultimate goal will be to obtain enough working information to develop a computer model of the structure/environment interaction and look at the potential failure mechanics. This evaluation program becomes an important part of the restoration work especially if the decision is made to remove the steel frame surrounding the structure.

The stabilization and restoration efforts on the aqueduct are expected to take several years. The ASCE/NCS evaluation program is intended to be a permanent part of the aqueduct's future. The instruments are unobtrusive and are designed to fit into the historic makeup of the site. Interpretive signage will be placed on the structure by the NPS to inform the public about the engineering efforts and the data being collected and its support of the structural and architectural restoration efforts. If successful, lessons learned from this behavior evaluation program will be applied to other historic structures in the C&O Canal National Historic Park. ASCE/NCS is committed to maintaining its role as an active partner in the preservation of the canal system and its historic structures.

HISTORIC TRUST

On June 15, 1998 the Monocacy Aqueduct was included on the National Trust for Historic Preservation's ***1998 List of Endangered Historic Structures***. The aqueduct served as a monumental background for the ceremonial announcement of this year's list (see Photo 3). ASCE NCS representatives and ASCE's District V Director, Mr. J. A. Padgett participated in the event, explaining our monitoring program, answering questions on the aqueduct and its history, displaying monitoring instruments, and allowing the public to review the structure through a surveying instrument (see Photo 4). First Lady Hillary Rodham Clinton was among the dignitaries at the event, and she acknowledge ASCE's efforts in preserving the aqueduct in her speech.

Photo 1: Monocacy Aqueduct, looking northeast. The steel harness applied in 1975 to stabilize the structure is evident on the masonry arches and piers.

Photo 2: Field Work a) Author Steve Pennington on aqueduct sighting through transit in vertical survey grid. Note the harness tendon rods across aqueduct prism to stabilize structure. b) Author Bernie Dennis holding measured rod for horizontal survey grid.

Photo 3: First Lady Hillary Rodham Clinton (center) receives tour of aqueduct site at June 15, 1998 ceremony. NPS C&O Canal Superintendent Doug Faris (in Park Service uniform to left of Mrs. Clinton) explains ongoing efforts to dignitaries including Maryland Congressman Roscoe Bartlett (left of Superintendent), former Maryland Congressman Gilbert Gude (right and behind Mrs. Clinton), Maryland Senators Paul Sarbanes (immediate right of Mrs. Clinton) and Barbara Mikulski (far right).

Photo 4: Authors Bruce Mattheiss (left) and Bernie Dennis (right) attend ASCE table at the June 15th event naming the Monocacy Aqueduct on the 1998 List of Endangered Historic Structures announced by the National Trust for Historic Preservation.

The Niagara Gorge Bridges

Carl J. Lehman, P.E., F.ASCE [1]

Abstract

The Niagara River establishes the international border between New York State, United States and Ontario Province, Canada. Niagara Falls, famous as the waterfall which conveys the world's largest volume of water, lies along this river. The portion of the Niagara River below the Falls and extending to Lake Ontario is known as the Niagara Gorge. Since 1848 many bridges of various styles have spanned the gorge below the Falls. The earliest structures were suspension bridges. These cable supported structures were gradually replaced with steel arches which proved to be the most compatible structural form for these environs. In April 1992, the ASCE designated the "Bridges of Niagara" an International Historic Civil Engineering Landmark. This paper summarizes the history and evolution of the bridges which have crossed this historic site and describes those currently in service to the international community.

Introduction

Plans for a bridge across the Niagara gorge were considered as early as 1824. These plans placed the site of a structure crossing the Niagara River between Queenston, Ontario, Canada and Lewiston, New York, United States. Discussion of a bridge spanning the gorge immediately below the Falls began around 1845. Charters were established in 1846 by New York State and the Canadian Government for construction of a bridge across the gorge. Although the scope of the bridge was not determined at this point, the charters recognized the probable future demand for a bridge across the gorge. Leading engineers of that time who considered a span at this location feasible were Charles Ellet, Jr., John Augustus Roebling, Edward W. Serrell, and Samuel Keefer. All four of these men would eventually build a

---

[1] Bridge Inspection Team Leader, Hardesty & Hanover, L.L.P., 1501 Broadway, New York, NY 10036

suspension bridge across the gorge between 1848 and 1869. Several other engineers would follow with ever improving designs founded on past shortcomings. Successful bridges would be designed and constructed by Leffert L. Buck, Charles C. Schneider, Waddell & Hardesty, and Hardesty & Hanover.

First Suspension Bridge

Charles Ellet, Jr. was the first engineer to successfully erect a simple, yet effective, suspension bridge across the gorge. The first wire cable, consisting of thirty-six No. 9 wires, was strung between the United States and Canada in early 1848. A light iron basket was suspended from two iron pulleys which ran on top of the cable. A wooden plank platform for carrying materials and tools was suspended from the pulleys below the iron basket. This "basket ferry" was used for nearly a year by construction workers and public passengers.

In July of 1848, Ellet completed a temporary wooden plank suspension bridge to be used during construction of a proposed Railway Suspension Bridge on that site. This bridge had a width of approximately eight feet and was supported by wooden towers at each bank. This bridge was the completion of Ellet's work at the gorge.

Lewiston & Queenston Suspension Bridge

Construction of the first permanent bridge over the gorge began in 1850 under the direction of engineer Edward W. Serrell. Opened on March 20, 1851, the Lewiston & Queenston Suspension Bridge was proclaimed to be the longest suspension bridge in the world at that time. Small stone towers were erected on rock plateaus higher than the main bridge deck. The span between the towers was recorded to be 1,040 feet although the deck below only measured 849 feet in length. A severe windstorm in 1864 destroyed this bridge.

When the electric railroads became established nearly 35 years later on both sides of the river, it was decided to rebuild the bridge to connect the railroads. The replacement suspension bridge was designed by engineer Richard S. Buck and construction was completed in July of 1899. This structure remained in service until November of 1962 when the Lewiston-Queenston steel arch bridge was opened. The suspension bridge was closed and was eventually demolished.

Niagara Railway Suspension Bridge

Prior to construction of his Niagara Railway Suspension Bridge, John A. Roebling had built six suspension bridges primarily for light highway traffic. Specifications for his Railway Suspension Bridge, on the site of Ellet's suspension bridge, proposed the use of a double deck to separate the trains from the public roadway traffic. He reasoned that a dual use single deck would be unsafe since the steam locomotives would undoubtedly frighten passing horses. The roadway was designed to be carried below the railroad. This arrangement turned out to be advantageous to the structural capacity and stability of the bridge. The added depth of the stiffening trusses which connected the two decks provided the rigidity necessary to carry railroad traffic. The span of this bridge between towers was approximately 821 feet with a deck length of 800 feet. The stiffening truss depth was 18 feet, the lower roadway deck width was 24 feet and the railroad deck width was 25 feet. The first train crossed this bridge on March 5, 1855 with "no vibrations whatever..." noted by Roebling. The Niagara Railway Suspension Bridge will always be considered a monument to engineering skill.

Leffert L. Buck was retained to design and complete work for the renovation and repair of the Railway Suspension Bridge. In 1877, the outside layers of wires in the cables were found to be corroded at the anchorages. This deterioration was repaired and two new anchorages were added behind the original anchorages. This increased the total anchorage strength by approximately 50 percent. The next renovation involved the replacement of the stiffening trusses. The wooden trusses were replaced with metal trusses without traffic interruption in 1880. This change decreased the cable dead load by 178 tons and permitted a safe increase in live load from 200 to 350 tons. The final renovation was the renewal of the stone towers carrying the cables. The tower stones were slowly disintegrating due to the inferior quality of the stones. Defective stones were replaced with sound stones when needed but the rocking of the towers under live load and the destructive action of frost made continued in-kind repairs impractical. The stone towers were replaced with iron towers in 1886. This completed renovation of the suspension bridge. The only remaining parts of the original bridge were the cables, saddles, suspenders and anchorages.

Niagara Railway (Whirlpool Rapids) Arch Bridge

Ground was broken on April 9, 1896 for construction of the first steel arch bridge across the gorge designed by engineer Leffert L. Buck. The proposed structure was a two-hinged spandrel braced arch which was to be constructed on the centerline alignment of the Railway Suspension Bridge. This bridge was to be built below and around the existing suspension bridge and would eventually support the existing railway trackbed. Work was successfully planned and completed without interruption to railroad or highway traffic.

The skewbacks of the arch span were founded on the "Clinton Ledge" of solid grey limestone about halfway between the water and the top of the gorge. The arch trusses were erected by building out the two halves of the arch as cantilevers anchored to the solid rock on top of the bluffs. The cantilevers were suspended by adjustable anchor chains connected to the arch at the top of the end post. The span of the arch between skewbacks was 550 feet with a rise of 114 feet at the crown. Two short approach spans of 115 feet connected the arch from directly above the skewbacks to the bluffs at each side. All work on the bridge was completed on August 27, 1897.

This bridge is now known as the Whirlpool Rapids or Lower Steel Arch Bridge and is the only nineteenth-century bridge spanning the gorge which has survived essentially unaltered. This bridge remains in service to public passage.

Niagara Falls & Clifton (Upper) Suspension Bridge

Samuel Keefer began work on his suspension bridge in 1867 but operations were suspended in October. Over the winter an ice-bridge formed just below the Falls providing an opportunity for two carrier ropes to be taken across the river in February of 1868. Work resumed in May of 1868. Buggies suspended from pulleys running on the carrier ropes were used to ferry men and materials across the river, much like Ellet's basket ferry of 1848. The completed bridge was opened in January of 1869. The cables were supported on towers of white pine approximately 100 feet tall. These towers supported a cable span of 1,268 feet. This was the longest single span in the world until the Brooklyn Bridge was completed in 1883.

The bridge deck was only 10 feet wide and could carry a single carriage. After nearly twenty years of service, Leffert L. Buck was retained to reconstruct the suspension bridge. The wooden towers were replaced with iron towers in 1884, prior to Buck's work. New cables, anchorages and truss were erected in 1888, leaving nothing of the original construction. The reconstructed bridge deck was 17 feet wide. A violent storm on January 10, 1889 fractured the truss deck connections to the cables and the deck fell into the river. The bridge was immediately rebuilt only to be replaced shortly after by the Upper (Falls View) Arch Bridge.

Niagara Cantilever Railway Bridge

Charles C. Schneider was contacted in October of 1882 by the Central Bridge Works of Buffalo, New York requesting him to prepare a preliminary estimate for a double-track railroad bridge of 900 feet clear span. The proposed bridge site was approximately two miles downstream of the Falls and only a few hundred feet upstream from the Niagara Railway Suspension Bridge. After careful consideration of an approximate profile of the proposed bridge site, a counterbalanced cantilever plan was recommended as the most feasible and economical for that location. He specified that the entire structure be made of steel except for the floorbeams and stringers which would be wrought iron. The contract was awarded on April 11, 1883 with the stipulation that all work be completed by December 1st of that year. Since Schneider was uncertain about the availability of quality steel on such short notice, he limited his use of steel to pins and heavy compression members.

The total bridge length between abutment anchorages was approximately 910 feet with a distance of 496 feet between the centers of the pier towers. The approach span cantilevers were 207 feet in length with the main span cantilevers being 188 feet. The suspended span between the main span cantilevers measured 120 feet. The pier tower height was 132 feet with the cantilver truss depth being 56 feet on top of the towers. The bottom chords of the cantilevered trusses sloped upward away from the towers with the truss depth tapering to 26 feet at the suspended truss span and 21 feet at the abutments. The pier towers were supported on masonry pedestals of Queenston limestone which were 38 feet high. The transverse distance between pier towers was 28 feet and the proposed trackbed was approximately 240 feet above the water surface. The use of pin-connections made it possible to complete construction in just over eight months. The bridge was officially opened on December 20, 1883.

The cantilever bridge was strengthened in 1899 but eventually proved to be inadequate for the steadily increasing weight of trains. A two-hinged spandrel braced steel arch bridge was designed by H. Ibsen, special engineer for the Michigan Central Railroad with construction beginning in 1923. The bridge was located approximately 75 feet downstream from the Cantilever Bridge and has a main arch span of 640 feet. The replacement structure came into service on March 16, 1925. The Michigan Central Arch Bridge remains in service today and is now known as the Conrail Arch Bridge.

Upper (Falls View) Arch Bridge

Increased tourist traffic and the need to provide a connection for the electric railroads at the location of the Upper Suspension Bridge led to the construction of Leffert L. Buck's second steel arch bridge over the Niagara. He was retained for his expertise gained through the construction of the Niagara Railway Arch around the existing Railway Suspension Bridge in 1897. He designed a steel two-hinged parallel-chord trussed spandrel arch with a span of 840 feet between skewbacks and 42 feet between the arches. The arch rose to approximately 240 feet above the water and carried a roadway 46 feet wide. The main span was connected to the abutment anchorages by two simple truss approach spans of 210 feet. The total bridge length between abutment anchorages was 1,260 feet. This bridge was constructed around the existing suspension bridge with little to no interruption to traffic. Construction was completed in June of 1898. This bridge is recognized as the longest steel arch bridge built in the nineteenth century.

The skewbacks for this structure were located approximately 40 feet above mean water level near the base of the gorge walls. A record ice jam in the gorge rose to 50 feet above mean water level and extended above the level of the skewbacks. The slowly moving mass of ice sheared the arch at its hinges and the steel bridge collapsed into the gorge on January 27, 1938.

Rainbow Arch Bridge

Shortly after the collapse of the Falls View Bridge it was decided that a structure at this location was necessary for the economic health of the area. The firm of Waddell & Hardesty was selected to design a replacement structure. The new fixed-end twin steel girder arch bridge was to be located approximately 550 feet downstream from the previous bridge. Due to the history of ice jams in the gorge the skewbacks are located 55 feet back from the shoreline and 52 feet above mean water level. The span of the new bridge is 950 feet between skewbacks and rises to 204 feet above water level at the crown. The total bridge length between abutment anchorages is 1,430 feet. Ground was broken for the bridge on May 16, 1940 and the bridge was opened to traffic on November 1, 1941. This bridge remains in operation today serving the local residents and tourists of the Niagara Falls area.

Lewiston-Queenston Arch Bridge

The gradual increase in traffic and vehicle weights made it necessary to proceed with the construction of a steel girder arch bridge to replace the functionally obsolete Lewiston & Queenston Suspension Bridge. A new bridge was designed by Hardesty & Hanover, the successors to Waddell & Hardesty who had designed the highly successful Rainbow Arch Bridge. The Lewiston-Queenston Arch Bridge is located approximately seven-tenths of a mile upstream from the original suspension bridge with an arch girder span of 1,000 feet between skewbacks. The bridge was opened on November 1, 1962 and remains the world's longest fixed-end steel girder arch bridge. This bridge is the youngest of the structures across the gorge and continues to serve the travelling public today.

Summary

The design and construction of bridges across the Niagara Gorge was necessary to the historic and economic development of the Niagara Falls area. The demand for a safe and convenient connection between the United States and Canada was easily apparent. This need provided a great challenge to civil engineers and to the development of bridge engineering technology. The successful crossing of the gorge required the skill of many engineers willing to take risks and extend their engineering knowledge beyond established limits. Their foresight and intuition significantly contributed to the refinement and advancement of design techniques for suspension and arch bridges for all time.

References

Buck, Richard S. "The Niagara Railway Arch.", ASCE Transactions, Volume XL, No. 836, December 1898.

Greehhill, Ralph (1984), Spanning Niagara - The International Bridges 1848-1962, The University of Washington Press, Seattle & London.

Lehman, Carl J., "The Rainbow Bridge", ASCE Practice Periodical on Structural Design and Construction, Volume 1, No. 1, February 1996.

Roebling, John A. (1855), "Memoir of the Niagara Falls Suspension and Niagara Falls International Bridge."

Schneider, Charles C., "The Cantilever Bridge at Niagara Falls.", ASCE Transactions, Volume XIV, No. 317, November 1885.

Skinner, Frank W., "Types and Details of Bridge Construction.", *The Engineering Record*, Volume 48, No. 5, August 1, 1903 & No. 6, August 8, 1903.

Vidler, Virginia (1985), Niagara Falls - 100 Years of Souvenirs, North Country Books, Inc., Utica, New York.

"Rainbow Bridge, Niagara Falls.", *The Engineer*, November 21 & 28, 1941.

The City Plan of Philadelphia, a National Historic Civil Engineering Landmark

David Clifford Hanly, Associate Member, ASCE [1]

Abstract

The Board of Directors of the American Society of Civil Engineers voted in 1996 to designate the City Plan of Philadelphia as a National Historic Civil Engineering Landmark. The decision of the Board was based on the recommendation of the Committee on History and Heritage of American Civil Engineering. This paper describes the City Plan of Philadelphia, planned in 1683, and highlights the characteristics that make the plan notable as a work of civil engineering and of national historic significance. The key features of the plan are:

1. Gridiron street pattern
2. Street widths appropriate to the streets' functions (a "First" in American City Planning)
3. Open public squares (a "First" in American City Planning)
4. Central public square for future public building
5. The planners' foresight in providing ample land for future development within the plan (a "First" in American City Planning)

Introduction

The City Plan of Philadelphia is a seminal creation in American city planning. It was the first American city plan to provide open public squares for the free enjoyment on the community and a gridiron street pattern featuring streets of varying widths: wide main streets and narrower side streets. Additionally, the City Plan of Philadelphia was the first city plan in the United States to provide for long-term urban growth. These features inspired the planners of many cities to adopt the Philadelphia Plan as a model.

The Board of Directors of the American Society of Civil Engineers voted in 1996 to designate the City Plan of Philadelphia as a National Historic Civil Engineering

----

[1]Struct. Engr., Frederic R. Harris, Inc., 260 S. Broad St., Phila., PA 19102, USA.

Landmark. The decision of the Board was based on the recommendation of the Committee on History and Heritage of American Civil Engineering. The Society dedicated a commemorative plaque on Logan Circle in Philadelphia on July 18, 1997.

In order to fully appreciate the final plan it is vital to know how the plan developed. The city's final design developed from a grand vision, took shape through consideration of alternatives and finally came to fruition by exploiting the potential of the local site conditions. In short, the final plan developed in the same way as do all great civil engineering projects. The following is a short history of the planning of the City of Philadelphia.

The Grand Vision

In 1681 Charles II granted William Penn the Province of Pennsylvania, a region nearly the size of England. It is not surprising that Penn had a grand vision considering the grand scale of the province.

Penn founded the City of Philadelphia as the Capitol of the Province of Pennsylvania. Prior to this, there was not a group settlement at the site. Penn's plan for the Province of Pennsylvania called for purchasers of large farming tracts to receive proportional lots within his proposed Capitol. The farming tracts were located in the region surrounding the Capitol site. This initial concept is detailed in the document titled "Certain conditions and concessions agreed upon by William Penn, Proprietor and Governor of the Province of Pennsylvania, and those who are the adventurers and purchasers in the said province, the 11th of July, 1681" (Hazard 1850). This regional plan is similar to New England's compact, or square, agricultural villages. The New England plans located farmers' residences in the town and their fields in the surrounding country. Penn's plan took the idea one step further. He provided the capitol as a major commercial center and surrounded it with satellite agricultural communities.

That Penn was planning in recognizably "civil engineering" terms is clear from this excerpt from the "conditions and concessions":

> That so soon as it pleaseth God that the above persons arrive there, a certain quantity of land or ground plat shall be laid out for a large town or city, in the most convenient place upon the river for health and navigation; and every purchaser and adventurer shall, by lot, have so much land therein as will answer to the proportion which he hath bought or taken up upon rent But it is to be noted, that the surveyors shall consider what roads or highways will be necessary to the cities, towns, or through the lands. Great roads from city to city not to contain less than forty feet in breadth, shall be first laid out and declared to be for highways, before the

> dividend of acres be laid out for the purchaser, and the like observation to be had for the streets in the towns and cities, that there may be convenient roads and streets preserved, not to be encroached upon by any planter or builder, that none may build irregularly, to the damage of another. In this custom governs.

This excerpt shows that sufficient means of transportation were clearly forward in Penn's mind. He planned to provide access by ship to the Capitol City, dedicated right-of-ways within the city and towns and highways among towns and the city.

Hazard (1850) is an important source of documentation of the founding of Pennsylvania. It reprints many letters written at the time and most of the critical documents including the Charter granting Penn proprietary powers and the "conditions and concessions" between Penn and his first purchasers. The accuracy and veracity of Hazard is attested to by the fact that it is the legal source of case law for the period. (It should be noted that the spellings of place names vary in the original documents and that these spellings are reprinted here without comment.) Hazard (1850) reprints instructions written by Penn to his "commissioners for the settling of the present colony." These instructions expanded on the initial criteria (quoted above) for siting the Capitol City. The commissioners were John Bezar, Nathaniel Allen, William Haig and William Crispin (who died shipboard). A pertinent excerpt follows.

> That having taken what care you can for the people's food, in the respects abovesaid, let the rivers and creeks be sounded on my side of Delaware River, especially Upland, in order to settle a great town, and be sure to make your choice where it is most navigable, high, dry and healthy; that is, where most ships may best ride, of deepest draught of water, it possible to load or unload at the bank or key side, without boating and lightering of it. It would do well if the river coming into that creek be navigable, at least for boats, up into the country, and that the situation be high, at least dry and sound, and not swampy which is best known by digging up two or three earths, and seeing the bottom.

Here Penn specifies an early Percolation Test. Although his specification is not too rigorous, it does show Penn's scientific bent. The intention is clearly to locate the city at a healthy spot with an eye toward the oldest concerns of civil engineering: Public Health and Sanitation. These were the enemies of all new colonies. Colonies that were planted in low tidal land were subject to illness due to contaminated drinking water and malaria.

Penn's agreement with the first investors and instructions to the commissioners demonstrate the grand vision for the Capitol City. He planned a city located on high ground with a vibrant port and efficient roadway access to the countryside.

Consideration of Alternatives

Upon their arrival in the province, the commissioners met with Deputy Governor William Markham, who was already in the Province. Together they investigated potential sites on the west bank of the Delaware River. Upland, the only existing settlement of size, was considered for the capitol. In the spring of 1682 the present site of Philadelphia was selected between the Schuylkill and Delaware Rivers.

Exploiting the Potential of the Local Site Conditions

Thomas Holme, Surveyor-General of the Province, arrived at Upland on August 3, 1682 with instructions to survey the city lots of purchasers who had arrived in the Province. A preliminary drawing for lots was made by Markham, Holme and the commissioners September 19, 1682. This selection, later voided by Penn, set lots on four streets parallel and adjacent to the Delaware River bank. This assignment is inconsistent with Penn's prior instructions to the commissioners directing that all of the lots front on the river. It appears that the commissioners, probably in consultation with Markham and Holme, modified the plan to better suit actual conditions after the selection of the site. At this stage the final plan had not yet been established and was evolving in the minds of the men on the site.

William Penn arrived in his Province for the first time on October 27, 1682. He approved the Philadelphia site but brought with him the updated list of purchasers, including major deletions and additions, which invalidated the previous month's lot selection. It appears that it was at this stage that Penn directed that land be allotted on both the Schuylkill and Delaware Rivers. Who would have anticipated the propitious site selected by the advance team? Not only was the site high and dry, as was directed, but also fronted on two rivers. Clearly the settlement plan evolved based on the local terrain.

In the end, Penn, Markham, Holme and the commissioners established the city plan of Philadelphia as it appears in the famous "Portraiture of the City of Philadelphia." This engraving of the city plan was published in London in 1683 to advertise for new emigrants. "Thomas Holme Surveyor-General" appears in the heading on the engraving. Holme wrote a letter dated August 16, 1683 that accompanied the Portraiture. The heading on the letter reads "A Short Advertisement upon the Scituation and Extent of the City of Philadelphia and the Ensuing Plat-form thereof, by the Surveyor General". The following is a portion of the letter as reprinted in Myers (1912).

> The City of Philadelphia, now extends in Length, from River to River, two Miles, and in Breadth near a Mile; and the Governour, as a further manifestation of his Kindness to the Purchasers, hath freely given them

> their respective Lots in the City, without defalcation of any their Quantities of purchased Lands; and as its now placed and modeled between two Navigable Rivers upon a Neck of Land, and that Ships may ride in good Anchorage, in six or eight Fathom Water in both Rivers, close to the City, and the Land of the City level, dry and wholesome: such a Scituation is scarce to be parallel'd....
>
> The City is so ordered now by the Governour's Care and Prudence, that it hath a Front to each River, one half at Delaware, the other at Skulkill.... The City, (as the Model shews [sic]) consists of a large Front-street to each River, and a High-street (near the middle) from Front (or River) to Front, of one hundred Foot broad, and a Broad-street in the middle of the City, from side to side, of the like breadth. In the Center of the City is a Square of ten Acres; at each Angle are to be Houses for publick Affairs, as a Meeting-House, Assembly or State-House, Market-House, School-House, and several other Buildings for Publick Concerns. There are also in each Quarter of the City a Square of eight Acres, to be for the like Uses, as the Moore-fields in London; and eight Streets, (besides the High-street, that run from Front to Front, and twenty Streets, (besides the Broad-street) that run cross the City, from side to side; all these Streets are of fifty Foot breadth.

The "Moore-fields" referred to were an open public park outside the walls of the old-city section of London. It was a wetland area which was drained in 1527 and laid out with walkways in 1606. It was developed not long after the founding of the City of Philadelphia.

Penn provided an initial grand vision that took its final form in the City Plan of Philadelphia. After consideration of alternative sites, the neck of land was selected. The planners modified the initial concept to enjoy frontage on two rivers.

Colonial City Planning, a Comparison

Before describing the Philadelphia City Plan in detail, it is useful to provide a brief description of selected contemporary colonial cities for comparison purposes. The following cities are recognized as having the important city plans of Colonial America (Reps 1965). European immigrants to the New World have entered through these cities for 375 years and, naturally, the plans of most of the cities of the North American interior were shaped by the impressions left by these plans.

<u>New Haven, Connecticut</u> - Founded in 1638, the city plan of New Haven is a prime example of the New England compact, or square, agricultural village plan. The plan divides a square town into nine blocks of equal size (a

gridiron pattern of four longitudinal and four transverse streets.) The central block serves as a large, square village green at the heart of the settlement. Residential and commercial buildings occupy the remaining squares. The village green became the site of public buildings and the market.

Annapolis, Maryland - The 1695 French baroque-style city plan of Annapolis was conceived by Francis Nicholson, Governor of the Maryland colony. Influenced by Christopher Wren's monumental plan for the rebuilding of post-fire London, the ambitious plan locates major civic buildings on the highest points and surrounds them with circular parks. The major avenues bisect these circles, providing splendid views. This was the first use of the baroque style in American city planning.

Williamsburg, Virginia - Established as the new colonial capital in 1699, Williamsburg was planned as a seat of government by Francis Nicholson, who was then Governor of Virginia. Drawing on his experience in laying out Annapolis and without the benefit of varied topography, Nicholson planned a town with a gridiron street pattern with the ends of the major thoroughfares terminating at significant buildings: the Capitol building, the College of William and Mary, and the Governor's Palace. This baroque city plan creates a charming closed town.

Savannah, Georgia - The Savannah city plan is recognized as a National Historic Civil Engineering Landmark for its influential repetitive modular grid. The city was sited and laid out in 1733 by James Oglethorpe, a trustee of the Colony of Georgia.

Charleston, South Carolina - In 1717 a city plan (conceived prior to 1680) was superimposed on the existing town. The plan incorporates a 0.8 hectare (2 acre) public square at the intersection of the two principal streets. Due to the small area reserved, the effect was to reserve the four corners at the intersection for public buildings. During the eighteenth century the four corners became occupied by the town market, a church, an arsenal and the courthouse, eliminating any open public space.

The following cities do not have carefully considered plans in the way that the preceding have, but are included due to their prominence in Colonial America.

New Amsterdam / New York City - The first permanent European settlement on lower Manhattan Island was established in 1625 consisting primarily of a fort owned by the Dutch West India Company. The settlement grew without a plan, and with natural irregularities. The English took possession in 1664 and the city government continued to survey plots and streets when required and without a comprehensive plan. This ad hoc development continued until 1811.

Boston, Massachusetts - The Massachusetts Bay Colony moved its principal town to the site of Boston in 1630 for good civil engineering reasons: potable water and a good harbor. As is evidenced by the modern city, Boston grew by necessity and tradition, not by imposed planning. Allowing for irregularities of topography, the principal streets radiate from the harbor to the countryside or lie tangential to the harbor.

It is important to also mention the reconstruction of London following the Great Fire of 1666. Not surprisingly, discussion of the city reconstruction plans turned all educated Londoners of the period into armchair Planners. It is likely that the plan proposed by Richard Newcourt is reflected in the Philadelphia plan (Reps 1965). Newcourt proposed five public squares, one central and four quadrant squares. Although Newcourt's plan was widely discussed, the plan of the great architect Christopher Wren was eventually selected and constructed in London.

Description of the City Plan of Philadelphia

The characteristics which distinguish the City Plan of Philadelphia, all "firsts" in American city planning are:

1. Street widths appropriate to the streets' functions
2. Open public squares
3. Foresight to provide ample land for future development within the plan

Street Widths Appropriate to the Streets' Functions

The Philadelphia city plan features an advance in the use of the gridiron street pattern in its provision of appropriately non-uniform street widths. This marked a first in the history of American city planning. The principal axis streets were 30 m (100 ft.) wide, streets fronting the rivers were 18 m (60 ft.) wide and the remaining streets were 15 m (50 ft.) wide. These widths can be seen graphically in Thomas Holme's 1683 "Portraiture" of the city plan and in the modern map of the city. The curb-to-curb dimensions established in the Original Plan are still used today. These dimensions were extravagantly generous in their day and adequately served the city until the advent of the automobile.

Open Public Squares

The open public squares are the most striking feature of the city plan. While the gridiron street plan had been used in the American colonies as early as 1638 in New Haven, the only public squares provided in earlier colonial plans were for the expressed use of civic buildings and markets. William Penn specified that the Philadelphia squares, other than Center Square, were to remain open for the recreation of the residents. As is mentioned above, this part of Philadelphia's city plan might derive from Richard Newcourt's plan for the reconstruction of London following the Great Fire in 1666. If this is the case, then the Philadelphia Plan is the first to put a wonderful idea to use.

The four quadrant public squares, one in each quarter of the original grid pattern, currently serve city residents as public parks. They provide unique character to their neighborhoods and a retreat for the citizens just as the planners envisioned. While none of the squares has continuously served this purpose, each has evolved with the neighborhood that surrounds it, providing a convenient, open public space when required.

Penn designated the *Center Square* as the site of future public buildings. Central public squares, as has been noted, appear in city plans prior to the Philadelphia plan. The New England compact, or square, agricultural town plan (of which New Haven is an example) provided a relatively large center public square. In this way Philadelphia is an important development which incorporates this earlier, successful planning aspect. Center Square's first public service in this capacity was as the site of Philadelphia's first municipal water supply, opened in 1801. This waterworks itself is distinguished as a National Historic Civil Engineering Landmark as the first major municipal waterworks to employ steam powered pumping methods in the United States. When the size of the city finally caught up with Penn's original vision, the city fathers moved the city offices to the square at its center. City Hall, as the municipal building is called, is a marvel in itself and a monument to William Penn. It took thirty years to construct (1870-1900) and occupies 1.8 hectares (4.5 acres). At a height of 167 m (547 ft.), it is the tallest masonry-bearing building in the world. Atop the City Hall tower stands an 11 m (36 ft.) high, 58 kN (26 ton) bronze statue of William Penn himself.

Foresight to Provide Ample Land for Future Development within the Plan

Thomas Holme's letter which accompanied the Portraiture implies that it was Penn's foresight to provide ample land for future development within the plan. Holme writes, "The city is so ordered now, by the Governour's Care and Prudence, that it hath a Front to each River, one half at Delaware, the other at Skulkill...." Penn provided a comprehensive plan for 5.2 sq. km (2 sq. mi.) of land in the wilderness (in 1681 there were, at most, a dozen residents) including a 4 hectare (10 acre) site for public buildings located 1.6 km (1 mi.) inland and a 3.2 hectare (8 acre) public park in a quarter of the city that would not be developed in the first hundred years. This long range planning is in contrast to the city plans cited above that either planned for the settlers who were in the ships at the dock and no more (New Haven), had no comprehensive plan (New York and Boston) or provided for the needs of the immediate colonial town (Annapolis and Williamsburg). The Charleston settlement was conceived with a gridiron street pattern but the plan was weakly implemented and the result is not distinguishable from the "unplanned" layout of New York. The modular grid system of the city plan of Savannah, founded half a century after Philadelphia, provided a pattern for future settlement. ASCE rightly designated the Savannah plan as a National Historic Civil Engineering Landmark, setting a precedent for the recognition of City Plans as landmarks. Foresight is the hallmark of a great plan and Philadelphia's plan was the first American plan to demonstrate the importance of foresight.

As mentioned above, Holme and Markham drew up a preliminary plan that located plots only on the Delaware side. But when Penn arrived, he ordered that the entire 3.2 km (2 mi.) width between the rivers and 1.6 km (1 mi.) in breadth be reserved for the city. It took 150 years to populate the entire 5.2 sq. km (2 sq. mi.) grid. That Penn's successors remained faithful to the gridiron street pattern and five public squares is a testament to the plan itself; the city grew into an early industrial-age center while remaining faithful to a plan laid out in the woods.

Conclusion

The City Plan of Philadelphia provided American city planners with a new model for the use of a gridiron street pattern featuring open public squares and street widths appropriate to the function of each avenue. The Plan also provided city planners with an example of long-range planning and the benefits which can be derived from it.

The city-scape of much of the nation can be traced to the City Plan of Philadelphia. The complete "Five Square" plan is used as a model by the city plans of several towns and in the cities of Raleigh, North Carolina and Tallahassee, Florida (Reps 1965). The landmark Philadelphia Plan (with at least one open public square) appears throughout the midwest and south, spreading mushroom-like from its origin in Philadelphia.

Appendix. References

Hazard, S. B. (1850). *Annals of Pennsylvania, from the Discovery of the Delaware, 1609-1682.* Hazard and Mitchell, Philadelphia, Pa.

Myers, A. C. (1912). *Narratives of Early Pennsylvania , West New Jersey and Delaware, 1630-1707*. Charles Scribner's Sons, New York, N.Y.

Reps, J. W. (1965). *The Making of Urban America; a history of city planning in the United States*. Princeton University Press, Princeton, N.J.

Subject Index

Page number refers to the first page of paper

Author Index

Page number refers to the first page of paper